Waste and Waste Management

Waste and Waste Management

Nuclear Waste: Safety, Security and Clean-Up
Ken Ykstra (Editor)
2024. ISBN: 979-8-89113-719-6 (Hardcover)
2024. ISBN: 979-8-89113-766-0 (eBook)

Enhancing the Use of Products with Recycled Contents in the Australian Construction Industry
Salman Shooshtarian, Tayyab Maqsood, Atiq Zaman, Savindi Caldera, Tim Ryley, Peter SP Wong
2024. ISBN: 979-8-89113-634-2 (Hardcover)
2024. ISBN: 979-8-89113-705-9 (eBook)

A Closer Look at Anaerobic Digestion
Rajesh Banu Jeyakumar, Kavitha Sankarapandian, Yukesh Kannah Ravi
2024. ISBN: 979-8-89113-358-7 (eBook)

Research and Development of Plastics Recycling Technology
Jeffrey R. Brady (Editor)
2024. ISBN: 979-8-89113-368-6 (Hardcover)
2024. ISBN: 979-8-89113-397-6 (eBook)

Ecology: Welcome Home
Algis Mickunas, Zilvinas Svigaris
2024. ISBN: 979-8-89113-222-1 (Hardcover)
2024. ISBN: 979-8-89113-262-7 (eBook)

More information about this series can be found at https://novapublishers.com/product-category/series/waste-and-waste-management/

Peter Clements
Editor

Agro-Industrial Wastes

Management, Valorization and Environmental Impact

Library of Congress Cataloging-in-Publication Data

ISBN: 979-8-89113-688-5 (softcover)
ISBN: 979-8-89113-788-2 (e-book)

Published by Nova Science Publishers, Inc. † New York

Contents

Preface

This book contains seven chapters on agro-industrial wastes. Chapter one delves into the challenges and opportunities associated with the management of agro-industrial waste. Chapter two highlights the relative composition of mushroom by-products and their nutritional value. Chapter three reviews topics such as the structure and composition of the eggshell, the recovery of eggshells as a source of CaCO3, and the mechanisms for obtaining CaCO3 nanoparticles (NPs) through nanotechnological processes, as well as the characterization of the NPs obtained from this biomaterial, along with its uses and applications. Chapter four is an analysis to determine the economic feasibility of using Red Tilapia (Oreochromis spp.) viscera for extracting oil and producing dry chemical silage (DCS). Chapter five proposes to determine the environmental impact of the use of chemical silage of red tilapia (Oreochromis spp.) viscera as a source of protein in the preparation of food for broilers (Gallus gallus domesticus) of the Ross 308 line, which have demonstrated their effectiveness without altering the productive variables of the foods designed. Chapter six shows that using chemical silage from red tilapia viscera in laying hens' feed is an economically feasible alternative in the constitution of a fish and poultry production system that generates higher profits than conventional production systems. Chapter seven focuses on determining the environmental impact of utilizing red tilapia (Oreochromis Spp.) viscera for producing chemical silage and incorporating it into the diet of laying hens, using the ecological footprint methodology as a sustainability indicator.

Chapter 1

Sustainable Solutions for Agro-Industrial Waste: A Holistic Perspective

Walter Murillo-Arango*, PhD
and Diego Fernando Montoya-Yepes, PhD
Chemistry Department, Faculty of Science University of Tolima, Ibagué, Tolima, Colombia

Abstract

This chapter delves into the challenges and opportunities associated with the management of agro-industrial waste. The contemporary waste management landscape grapples with complex issues driven by the rapid expansion of urban areas, industrial activities, and population growth, resulting in an intensified generation of diverse waste streams. Specifically, the chapter focuses on agro-industrial waste originating from agricultural processes. The discussion commences by highlighting the limitations inherent in traditional waste disposal methods, emphasizing the urgency of embracing sustainable alternatives for adequate valorization. Addressing environmental concerns, the chapter considers some impacts and proposes innovative solutions aligned with relevant sustainable development goals to mitigate escalating environmental degradation. Particular attention is devoted to unraveling the intricate interplay between agro-industrial waste and global environmental challenges, with a specific focus on its contributory role in climate change. Within this context, the chapter explores various alternatives for the valorization of agro-industrial waste. It delves into state-of-the-art technologies and methodologies designed to extract value from these waste streams, transforming materials previously

* Corresponding Author's Email: wmurillo@ut.edu.co.

In: Title Agro-Industrial Wastes
Editor: Peter Clements
ISBN: 979-8-89113-688-5

considered as refuse into valuable resources. The discussion investigates how these strategies not only yield economic benefits but also align with core principles of environmental sustainability. The chapter contemplates the management of agro-industrial waste through the integration of circular economy principles and places a significant focus on biotechnological pathways due to their potential applications at different scales. It advocates for a holistic and collaborative approach to address the challenges inherent in agro-industrial waste, encouraging the integration of sustainable practices into policy frameworks, agro-industrial operations, and individual behaviors.

Keywords: agro-industrial waste, waste management, waste valorization, environmental impacts, biotechnology

Introduction

Agro-industrial wastes refer to biomass primarily generated through the processing and handling of organic products from animals, plant cultivation, and the processing of fruits and vegetables. These wastes are produced throughout the food production and commercialization chain, including activities such as agriculture, livestock farming, fishing, and the transformation of agricultural raw materials into final products. The increase in population, urbanization, and globalization are among the main causes of food waste and the rise in agro-industrial waste generation. The significance of this issue is reflected in the substantial scientific literature produced on the subject. An example is clear when assessing the evolution of the number of records on the topic in the Web of Science, using keywords such as Agro-industrial waste, Agro-industrial Residues, Agro-industrial By-products, and Agro-waste. A search conducted on January 11 of the current year yielded 6278 records. Figure 1 illustrates the increasing production of related literature from 1997 to 2023.

Despite the growing interest in researching the topic, there are still gaps in specific figures about global food waste. However, according to the FAO's 2011 report, approximately one-third of the globally produced food is wasted, equivalent to 1600 million tons per year, with an economic loss of 1.2 trillion dollars. On a per capita basis, this waste amounts to around 74 kg/year (https://www.fao.org/fao-stories/article/en/c/1648308/). The issue is expected to escalate, with estimates suggesting that by 2030, annual food loss and waste will reach 2.1 billion tons, valued at 1.5 trillion dollars (Hegnsholt et al., 2018).

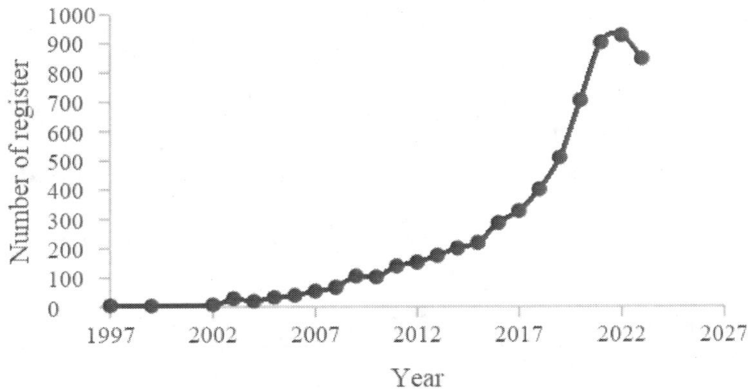

Figure 1. Temporal evolution of the number of records related to the agro-industrial waste topic in the WOS database.

The FAO distinguishes between food waste and food loss. Food waste relates to the reduction in quantity or quality due to actions by retailers, food services, and consumers. In contrast, food loss occurs during production, post-harvest, and processing, excluding retailers, food services, and consumers, based on decisions and actions by food providers in the chain (FAO, 2019). Analyses reveal a wide range of percentage losses at each stage of the supply chain, with higher levels in the case of fruits and vegetables, estimated between 45 and 60% according to FAO assessments (Bas-Bellver et al., 2020). The beverage industry incurs losses of 26%, followed by the dairy and ice cream industry at 21.3%, fruit and vegetable production and preservation at 14.8%, and cereal production at 30% (Capanoglu et al., 2022).

In the livestock sector, particularly concerning meat consumption, by-products are generated not only from the slaughtering process but also due to losses and waste of meat products throughout the supply chain. Losses in this sector have been increasing, mainly due to the growth in the global population, sometimes influenced by price decreases and shifts in consumption patterns. In 2023, pork production reached 122 million metric tons, chicken 140 million, beef 79 million, and sheep meat 17 million (Statista 2024). Like fruits and vegetables, meat products are perishable, leading to inevitable losses in the supply chain. Loss estimates for the sector globally stand at 20%, amounting to approximately 72 million tons in 2023.

Notably, these losses occur in the later stages of the food supply chain, generating waste primarily composed of organs, fat or lard, skin, feet, abdominal and intestinal contents, bones, and blood from cattle, pigs, and sheep, representing 66.0%, 52.0%, and 68.0% of live weight, respectively, along with dairy products such as cheese whey, curd, and milk sludge. Over half of the animal by-products are not suitable for normal consumption due to their unusual physical and chemical characteristics. Table 1 summarizes the percentage losses for various food products.

Table 1. Percentages of waste for different food products

Food Product	Food Loss and Waste (%)
Fruits and vegetables	45-60
Cereals	30
Fish and seafood	35
Meat	20
Dairy products	21,6
Beverage	26

The extent of this loss and waste, when viewed in terms of numbers, is alarming. Despite efforts to reduce it, driven by global organizations such as the FAO, controlling food waste is a complex task requiring the collaboration of different actors within the supply and commercialization chain, as well as a shift in perspective to view waste as valuable by-products that can support numerous transformative initiatives with economic, social, and environmental benefits. It is crucial to note that food waste is a consequence of three key global trends. Firstly, urbanization has extended the food supply chain, increasing the distance between production and final consumption. This entails transporting food over greater distances, necessitating better storage strategies to prevent losses. The second trend is the shift in diets, particularly in transitioning economies like Brazil, India, and China, where there is a move from starch-rich diets to those with more meat, fish, fruits, and vegetables, which are more perishable. The third trend is the growing globalization of trade and large-scale distribution, leading to increased demand for higher-quality products and safety standards, as well as a rise in the volume of traded products (Pinto et al., 2022). Changing these trends is likely to be complex due to the evolution of our social model, and aside from the inedible parts of food, food waste that cannot be avoided will continue to be generated, requiring proper management.

In terms of impact, food loss is intricately linked to food security, directly affecting food availability. It goes against principles of responsible production and consumption, promoting inefficient resource use, increasing pressure on agricultural systems, and generating significant environmental implications. This involves using natural resources for food production that is subsequently wasted, increasing the environmental footprint, and affecting both terrestrial and aquatic ecosystems, among other impacts. Discussing valorization alternatives within the framework of the circular economy and a bioeconomy that explores options for generating added value is urgent. In this context, agricultural by-products represent a valuable source of biomass that can be transformed from an inefficiently used and problematic resource into an asset contributing to the sustainability of various sectors. This chapter will address some elements related to the management, environmental impact, and valorization of agro-industrial waste, focusing on the recovery of active components and transformation, especially using biotechnological tools.

2. Impact of Food Loss and Waste

As previously mentioned, food loss is a global phenomenon linked to all stages of production and distribution chains—a problem that has become one of the greatest challenges currently faced by humanity (Awasthi et al., 2022). The impacts are not solely centered on global food security; the environmental consequences are evident, prominently visible in the compromised conditions and characteristics of aquatic and terrestrial ecosystems.

This is a critical point to consider as the global population continues to grow, and inadequately exploited or utilized natural resources become scarcer. Inefficient food management processes result in the loss of critical resources and contribute to the expansion of the environmental footprint. Therefore, concrete actions focusing on utilizing waste are necessary, starting with a comprehensive vision that considers environmental and social aspects, while also maintaining an economic focus (Sathish, Sasikumar, and Dhilipkumar, 2024).

In this context, the circular economy emerges as a relevant approach to optimize resource use and reduce pressure on ecosystems by closing production and consumption cycles (Khan and Ali, 2022). This involves implementing sustainable practices throughout the chains, aligning with the

Sustainable Development Goals (SDGs). It is essential to articulate the relationship between food loss and its environmental impacts, exploring specific indicators that measure and comprehend the magnitude of this challenge, serving as the foundation for different biomass valorization strategies.

2.1. Food Security

One of the most crucial and extensively addressed points in this context is food security, as food loss, beyond being an efficiency problem in supply chain processes, directly impacts the environment with global repercussions. According to the Food and Agriculture Organization (FAO), a person experiences food insecurity when lacking regular access to sufficient safe and nutritious food for normal growth, development, and an active and healthy life. This can result from insufficient food availability or a lack of resources to acquire it (FAO, 2023).

In this regard, production and distribution systems that do not work with comprehensive utilization of products and by-products incur losses, leading to environmental consequences ranging from resource depletion to ecosystem disruption. While food wastage implies inefficient use of nutritional resources—over 3.1 billion people worldwide (42%) could not afford a healthy diet in 2021 (FAO, et al., 2023)—it also contributes to increased pressure on natural resources, already undergoing aggressive exploitation processes, such as aquatic and terrestrial ecosystems.

SDG 2.1 emphasizes the need not only to eradicate hunger but also to ensure access to safe, nutritious, and sufficient food for all. The report "*The State of Food Security and Nutrition in the World*" presented by FAO, IFAD, and others in 2023, tracks dynamics related to food security. It highlights that, in SDG Indicator 2.1.1 related to global hunger, a prevalence of undernourishment of around 9.2% of the world's population was evident in 2022. For the same year, it was estimated that 29.6% of the world's population suffered from moderate to severe food insecurity, with 11.3% experiencing severe food insecurity (FAO, et al., 2023).

The connection between food loss or wastage and food security lies in the reduced availability of food for the population. Communities require a stable food supply for their daily sustenance, leading to increased vulnerability and instability in less developed territories. Additionally, competition for scarce resources can intensify, increasing pressure on agricultural systems and necessitating the expansion of agricultural frontiers, exacerbating soil degradation and water pollution.

In this sense, evaluating the effects of food loss on food security involves not only measuring the quantity of wasted food but also understanding how this phenomenon increases existing challenges in sustainable food provision. Effective solutions should not only address loss management but also adopt strategies that enhance the resilience of food systems to environmental pressures.

2.2. Pressure on Agricultural Productive Systems

The increase in the global population, coupled with climate change-induced extreme events, changes in land use, and the imminent reduction of available water in both volume and acceptable quality for agriculture, heightens the risks of not being able to produce enough food (Pérez, Leyva, and Gómez, 2018). The primary cause of the current global food crisis is not a production problem but rather an issue of inequality in access and distribution of food (Pérez, Leyva, and Gómez, 2018). According to FAO estimates, around 1.3 billion tons of food are wasted worldwide every year (Carretero, 2018), reflecting problems throughout the chain from production to final consumption. In 2018, Carretero pointed out that, in the European Union, households and processes of transformation, restoration, production, and sales wasted 53%, 19%, 12%, 11%, and 5% of the total volume of food, respectively.

This implies that waste in the production chain begins with production planning based on parameters different from effective food demand (Carretero, 2018). Additionally, technological level and economic capacity mean that, for example, low-income countries lack the infrastructure capacity for proper processing, transportation, and preservation of fresh foods, necessitating increased production to meet producer commitments, considering estimated losses of fresh products.

Trends like agroecology have suggested that while industrialized or extensive agriculture may increase food volume in the short term, it often operates under energy-inefficient models endangering resource sustainability (ODELA, 2019). This reality triggers ethical, economic, and social consequences, in addition to negative environmental consequences. Currently, global climate change is a phenomenon affecting not only production levels but also crop quality due to high temperatures, drought, flooding, and an increased incidence of pests and diseases. It is crucial to note that, from an environmental perspective, non-edible parts of food or discarded unsafe food remain relevant as waste requiring proper management schemes to minimize environmental impact, necessitating investment in technology and research that increases chain costs.

Consequently, some resources at risk, such as water use related to water footprint, are considerable. For example, producing one kilogram of beef requires around 15,000 liters of water. Additionally, the generated but unconsumed food occupies approximately 1.4 billion hectares of land, equivalent to almost 30% of the world's total agricultural areas (Carretero, 2018), promoting land and water concentration due to intensive resource use in these production models (ODELA, 2019), leading to serious social issues.

If food needs to increase by 70% by 2050, according to population growth trends, water availability would need to increase by 55%, and energy by 50% (FAO, 2011). The challenge is not only to achieve agricultural security, reducing negative externalities to the environment through utilization and valorization schemes, ensuring sufficient clean water, agricultural land, energy, and labor, allowing for food security (Pérez, Leyva, and Gómez, 2018).

2.3. Increase in Environmental Footprint

This problem not only affects food security, as highlighted, but also triggers significant environmental consequences, with the generation of greenhouse gases (GHGs) being one of the most concerning. Agriculture emits around a quarter of global greenhouse gases (FAO, 2019). In 2019, FAO estimated that about 14% of food, worth $400 billion, is lost after harvest and before reaching retail distributors. The total losses and waste of food were responsible for approximately 8% of greenhouse gas emissions at the time of the report.

Throughout the production chains, GHGs are released at every stage, depending on the process, contributing directly to climate change, and associated environmental, social, and economic impacts. CO_2 emissions, for example, are generated in food production, combined with discarded food transforming into waste that requires additional resources for management, resulting in impacts across various fronts. In the production phase, intensive agriculture and inefficient crop management are key factors in GHG generation. Food overproduction leads to deforestation to expand agricultural frontiers, releasing large amounts of carbon dioxide (CO_2) and preventing its capture by plants. FAO estimates that 3.3 gigatons of CO_2 are emitted from produced but unconsumed food; therefore, food waste would constitute the third-largest source of emissions globally, following the United States and China. Additionally, excessive use of fertilizers and pesticides contributes to the emission of nitrous oxides (N_2O), gases with a much higher global warming potential than CO_2.

As food progresses through the supply chain, from harvest to consumer, losses, and waste multiply. Food consumption and waste trends have been studied by tracking families in China, a densely populated country, based on the quantification of the home's ecological, carbon, and water footprint. It has been estimated that around 415 kg is the annual food consumption in a household, of which 1080 kg CO_2 eq of carbon, 673 m^3 of water, and 4956 gm^2 of ecological footprint. Waste (16 kg) alone is responsible for 40 kg CO2 eq of carbon, 18 m^3 of water, and 173 gm^2 of ecological footprint (Capanoglu, et al., 2022).

These decomposing foods in landfills emit methane (CH_4), a greenhouse gas with a heat-trapping capacity several times greater than CO_2. This process intensifies further when foods decompose under anaerobic conditions (Capanoglu, et al., 2022). Additionally, it can cause respiratory difficulties in humans and animals due to air pollution from dioxins, ash, and gases generated by waste incineration, contributing to groundwater contamination. Moreover, food waste dumping reduces its energy content; energy loss in landfills equals 43% of the energy provided for food preparation in the U.S. (Capanoglu, et al., 2022).

In addition to direct emissions of gases, it is crucial to consider that food loss also implies a waste of resources used in the production process, such as water, energy, and labor. This misuse of resources indirectly contributes to the environmental footprint of food production. It is noteworthy that about one-third of the world's soils are degraded due to overexploitation, releasing approximately 78 gigatons of carbon dioxide into the atmosphere, costing more than 10% of GDP due to biodiversity loss and ecosystem service depletion (FAO, 2019).

In this regard, it is essential to implement measures to reduce food loss throughout the supply chain. This will not only benefit global food security but also significantly mitigate greenhouse gas emissions associated with food production and disposal. Strategies such as more efficient harvest management, improved storage and distribution practices, valorization processes, and biomass utilization, along with public awareness of food waste, can play a crucial role in building a more sustainable and environmentally respectful food system. According to FAO, land restoration could eliminate up to 51 gigatons of carbon dioxide and increase food production by 17.6 megatons annually.

2.4. Indicators and Methodologies for Measurement

At this point, it is crucial to highlight that the lack of attention to this issue is closely related to the inadequate estimates of the magnitude of wasted food and, consequently, its environmental, social, and economic repercussions. As discussed by FAO in the "*Food Waste Index 2021*" report, very few governments have solid data on this issue that demonstrate the need for measures and prioritize efforts (UNEP, 2021).

This reality has led to an underestimation of generated waste, with suggestions that estimates developed by FAO could even double. This is considering the report on food loss and waste worldwide: scope, causes, and prevention, published in 2011 (Gustavsson, Cederberg, and Sonesson, 2011). An alarming fact is that only 9% of the global population lives in nations with reliable assessments of this phenomenon in households, and this percentage is equally low in retail, at 8%. However, it increases to 25% in the case of food services (UNEP, 2021).

Consequences environmental indicators that a state or organization could monitor, proposing appropriate indicators, include the release of greenhouse gases, water and energy consumption, land, and fertilizer use, as well as biodiversity decline (CEC n.d.). This could be estimated by considering indicators proposed by UNEP such as Indicator 12.3.1.a, the food loss index: measuring losses of major commodities in a country along the supply chain up to retail, excluding retail, and Indicator 12.3.1.b, food waste index: measuring food waste at the retail and consumer levels (in households or through food services) (UNEP, 2021). Both indicators are designed to achieve SDG target 12.3, focused on halving food waste and reducing food loss by 2030.

UNEP suggests working under a strategy based on setting goals, which should be monitored through constant measurements to finally propose and implement measures, termed "Target-Measure-Act." For this, it is necessary for member countries to submit reports periodically, clearly defining the scope, focusing on selected sectors. The methods used, while varying in territories, must be suitable and constantly monitored; ideally, studies should be conducted every two to four years to avoid measurement gaps and ensure reliable data (CEC n.d.; UNEP, 2021).

3. Strategies for Valorization

Throughout the discussion, the utilization of residual biomass from agro-industrial waste extends far beyond the food, energy, or fertilizer components. However, these avenues of utilization are the most straightforward, easily accessible, and generally require low technological investment. Nevertheless, attention in recent years has shifted towards the utilization of this residual biomass within circular economy frameworks, ensuring closed cycles and maximizing both the material and energy components of the substrate. This entails evaluating the possibility of transformations or recovery of nutrients and bioactive compounds before focusing on obtaining biogas and compost, aiming for the highest value and product diversification through the concept of biorefineries.

For example, the European Union defines bioeconomy in its strategy as the sustainable production of primary biomass and the conversion of organic resources (primary or residual) into food, feed, bioproducts, and bioenergy. It is considered a pillar linking various sectors such as agriculture, forestry, energy, waste management, the chemical industry, in addition to sustainable growth goals. In this scenario, biomass waste from primary, secondary, and tertiary sectors of economic activity is expected to play a significant role in supplying the raw materials needed for sustainable bioeconomy pathways.

The use and transformation of agro-industrial waste generally present lower environmental impacts, reducing competition for resources such as land for agricultural activities and ensuring food security. Additionally, it allows for the recovery of material and energy value from the waste, facilitating management within circular economy frameworks. The discussion around using energy crops for biofuel production has been complex and controversial. However, the use of agricultural residual biomass for energy production helps counteract these impacts, although technological development associated with the conversion of certain types of residual biomass, such as lignocellulose, remains a technological challenge.

An increasing amount of research focuses on applying different technologies, including physical and chemical pretreatments, extraction processes, and various bioprocesses involving biochemical transformations and biotechnological tools like fermentation and biotransformation. These aim not only at the recovery of compounds of interest but also at producing higher-value biomass for food purposes and obtaining a wide range of industrially relevant products.

3.1. Biorefineries Based on Agro-Industrial Wastes

The crisis in the oil industry, combined with international agreements on the environment, especially those stemming from the Kyoto Protocol, has led economic powerhouses like the United States, Japan, China, and several European countries to take regulatory, institutional, financial, and technological measures to promote the creation of biotechnological alternatives such as biorefineries in favor of the bioeconomy and environmental preservation.

In this context, a biorefinery can be defined as a transformation platform that converts biomass into different materials, products, or forms of energy with added value. These comprise integrated multiple productive structures optimizing input processing, with the main objective of increasing the economic potential of production processes by using by-products from production chains through the integration of different advanced conversion technologies for the generation of bio-based products.

It is crucial to note that from the perspective of sustainable development, alternative raw materials, chemicals, materials, and energy have become key challenges of this century. The use and transformation of agro-industrial waste play a crucial role in this context. Biorefineries of various types, from large industrial centers to small-scale facilities in rural areas, can be developed by complementing and incorporating existing processes to diversify the range of obtained products.

The biomass refining of agro-industrial waste for the development of commercial products, using sustainable and innovative bioprocesses, aligns closely with several goals established in the Sustainable Development Objectives. These include ensuring access to affordable, safe, sustainable, and modern energy; promoting inclusive and sustainable economic growth, employment, and decent work for all; building resilient infrastructure, promoting sustainable industrialization, and fostering innovation; making cities more inclusive, safe, resilient, and sustainable; ensuring sustainable consumption and production; adopting urgent measures to combat climate change and its effects; and sustainably managing forests, combating desertification, stopping and reversing land degradation, and halting biodiversity loss.

The processes of an agro-industrial waste biorefinery are classified into three main groups: biological pathway, thermochemical processes, and chemical processes. Biological pathways involve converting food waste into value-added products through enzymes or organisms, often using microorganisms. Thermochemical processes treat waste at high temperatures using liquefaction, pyrolysis, and gasification. Chemical processes involve the use of solvents for the extraction of components of interest, as well as the use of catalysts for transformation and valorization. The integrated combination of two or more of these processes has attracted attention due to its higher conversion efficiencies. Table 2 shows some examples.

Table 2. Examples of biorefining processes for certain residues and their transformations

Process type	Raw material	Methodology	Product	Reference
Biological	Corn cob	Microbial Fermentation with *Candida glabrata*	Bioethanol	(Shahid, et al., 2021)
	Ricotta Cheese Whey	Microbial Fermentation with *Cryptococcus laurentii*	Biodiesel	(Shahid, et al., 2021)
	sugarcane bagasse	Microbial Fermentation with *Rhodobacter capsulatus*	BioH_2	(Shahid, et al., 2021)
Thermo-chemical	rice husk	Pyrolysis with Microwave at 600W and 500°C	Nanoparticles and Carbon Nanotubes	(Debalina, Rajasekhar and Vinu 2017)
	Algae crops	Hydrothermal liquefaction	Biocrude	(Chen, et al., 2014)
	Urban solid waste	Gasification	Hydrogen	(Shahabuddin, and 2020)
Chemical	Beet greens	Pressurized liquid extraction with ethanol/water (70:30)	Phenolic compounds	(Ebrahumi, Mihaylova and Lante 2022)
	Green coffee beans	Extraction with subcritical water	Chlorogenic Acid	(Xu, Kim, and Choi 2019)
	Saffron processing waste	Pressurized liquid extraction with 1% citric acid pH 2.9	Anthocyanins and phenolic compounds	(Pappas, et al., 2021)

Around the world, numerous experiences have successfully consolidated competitive industrial-level biorefineries. These generate key products such as ethanol, biogas, and biodiesel. Examples include the Dupont project in Iowa, USA, producing 90,000 tons per year of ethanol and biogas from corn stover, and the Biochemtex project in Italy, producing 60,000 tons of ethanol and biogas per year from oats and Arundo donax.

In Latin America, Brazil stands out for its advanced technology in biorefinery operations, leveraging its experience in sugarcane exploitation for fuel production. The GranBio project in Alagoas, Brazil, produces 65,000 tons of ethanol and around 50 MW of electricity from sugarcane.

3.2. Chemical Composition of Residues

Data on the composition of different agro-industrial residues are variable, depending on factors such as the type of residue, its region of origin, and the production conditions. Notwithstanding these variations, understanding these values is crucial as they form the basis for establishing the most suitable valorization route.

Regarding the type of residue, authors like Cho et al. (2020) present figures on the quantity of biomass residues generated by agricultural and forestry activities, food processing, and other industrial sectors. Abundant biomass residues from agriculture include rice straw, wheat straw, corn stover, sugarcane bagasse, and rice husk, generating 731, 354, 204, 181, and 110 million tons annually, respectively.

Generated forestry residues are estimated at approximately 72.5 million tons in the United States and Canada alone. Other residues, such as rapeseed meal (35 Mt), citrus waste (15.6 Mt), banana waste (9 Mt), grape waste (5–9 Mt), and apple pomace (3–4.2 Mt), are generated worldwide each year. The olive oil industry produces around 30 Mt of residues annually, especially in Mediterranean regions. Coffee industry residues contribute 7.4 Mt of ground coffee and a substantial amount of coffee pulp, cherry husk, and silver skin. The use and utilization of such biomass are environmentally significant as agricultural by-products, including fruit and vegetable residues, lignocellulose, food waste, and slaughterhouse waste, are not adequately managed, leading to environmental issues. In Table 3, the chemical composition of some common biomass groups from agro-industrial residues is presented, highlighting the potential for value addition.

Table 3. Chemical composition of a selection of agro-industrial waste

Waste feedstock	Cellulose (%)	Hemicellulose (%)	Lignin (%)	Other regular components	Reference
Rice Straw	39.2	23.5	36.1	Phenolic compounds, flavonoids, and tannins.	(Sivakumar et al., 2022)
Wheat Straw	32	24	8.9	Phenolic compounds, terpenoids	(Sivakumar et al., 2022)
Barley Straw	33.8	21.9	13.8	Flavonoids	(Sivakumar et al., 2022)
Corn stalks	61.2	19.3	6.9	Terpenoids, alkaloids	(Sivakumar et al., 2022)
Cotton stalks	58.5	14.4	21.5	Phenolic compounds, terpenoids	(Sivakumar et al., 2022)
Potato peel	2.2	-	-	Phenolic compounds, Glycoalkaloids, Terpenoids	(Sivakumar et al., 2022)
Orange peel	9.21	10.5	0.84	Flavonoids, Terpenoids (essential oils), Carotenoids	(Sivakumar et al., 2022)
Pineapple peel	18.11		1.37	Phenolic compounds, Terpenoids, bromelain enzyme, Carotenoids	(Sivakumar et al., 2022)
Plantain peel	21	8	18	Phenolic compounds, Alkaloids, Flavonoids	(Hernández-Carmona, et al., 2017)
Plantain stalk	57.86	16.5	25.7	Phenolic compounds, Alkaloids	(Igbokwe et al., 2016)
Apple pomace	12	5.0	6.4	Phenolic compounds, pectin	(Sato et al., 2010)
Grape pomace	47	-	16.8	Phenolic compounds, pectin	(Spinei and Oroian, 2021).
Spend Coffee wastes	60	10	20	Lipids, proteins, phenols	(Cerino-Córdova, et al., 2020)
Sunflower stalks	42.1	29.7	13.4	Phenolic compounds, Terpenoids, Flavonoids	(Sivakumar et al., 2022)

These residues, besides containing significant amounts of lignocellulose, also harbor other metabolites of interest such as pectin, polyphenols, carotenes, especially in pulp and peels. Additionally, residues from oil extraction, like cake, are rich in lipids, proteins, fiber, and minerals, contributing to various uses in the food industry or the production of higher-value products. Table 4 provides a summary of the composition of several common oilseed cakes.

Table 4. Chemical composition of the cake of some common oilseeds

Oil Cake	Dry Matter (%)	Crude Protein (%)	Crude Fiber (%)	Ash (%)
Canola	90	33.9	9.7	6.2
Coconut	88.8	25.2	10.8	6.0
Cottonseed	94.3	40.3	15.7	6.8
Groundnut	92.6	49.5	5.3	4.5
Mustard	89.8	38.5	3.5	9.9
Olive	85.2	6.3	40.0	4.2
Palm Kernel	90.8	18.6	37	4.5
Sesame	83.2	35.6	7.6	11.8
Soybean	84.8	47.5	5.1	6.4
Sunflower	91	34.1	13.2	6.6

Adapted from: (Kolesárová, Hutňan, Bodík, and Špalková, 2011).

Residues from animal tissues like bones, fat, and muscle from poultry, bovines, and swine have compositions according to tissue type. Protein content is essential in both bones and muscles. These residues can be converted into biogas or composted, or undergo hydrolysis to extract proteins, enhancing their valorization. Resulting proteins can be used as ingredients in the food industry or to produce nutritional supplements. Bones, particularly rich in collagen, calcium, and phosphorus, can be used in animal feed or fertilizer industries. Fat content, more abundant in bovines and pigs than in chickens (Table 5), can find applications in the cosmetic, food, and biodiesel industries.

Table 5. Compositions according to tissue type

Animal tissue		Water (%)	Proteins (%)	Lipids (%)	Ash (%)
Fat	Poultry	28.7	3.7	67.4	0.3
	Bovine	4	1.5	94.0	0.5
	Pig	7.7	2.9	88.7	0.7
Bone	Poultry	51.0	19.0	9.0	15.0
	Bovine	46.0	19.0	15.0	20.0
	Pig	36.6	21.8	17.5	24.1
Muscle	Poultry	75.0	22.8	1.0	1.2
	Bovine	75.1	19.2	4.4	1.3
	Pig	75.1	22.8	1.2	1

Adapted from: (Pinto, et al 2022).

3.3. Extraction of Components of Interest

Extraction is the most critical step in obtaining bioactive compounds from agro-industrial residues. Phytochemicals such as phenols, alkaloids, terpenes, anthocyanins, among others, can be recovered from agricultural residues like roots, stems, pulp, peels, leaves, and seeds. These compounds are beneficial for treating human disorders and diseases due to their antioxidant properties, useful in preventing oxidative deterioration in food, cosmetic, and pharmaceutical matrices. Phenolic compounds, carotenoids, and anthocyanins are also used for their potential as natural colorants. Various extraction procedures are applied to isolate bioactive compounds, varying according to the type of compound of interest. Factors include the solvent type, temperature, particle size, tissue part used, and, in some cases, pressure, depending on the selected extraction method.

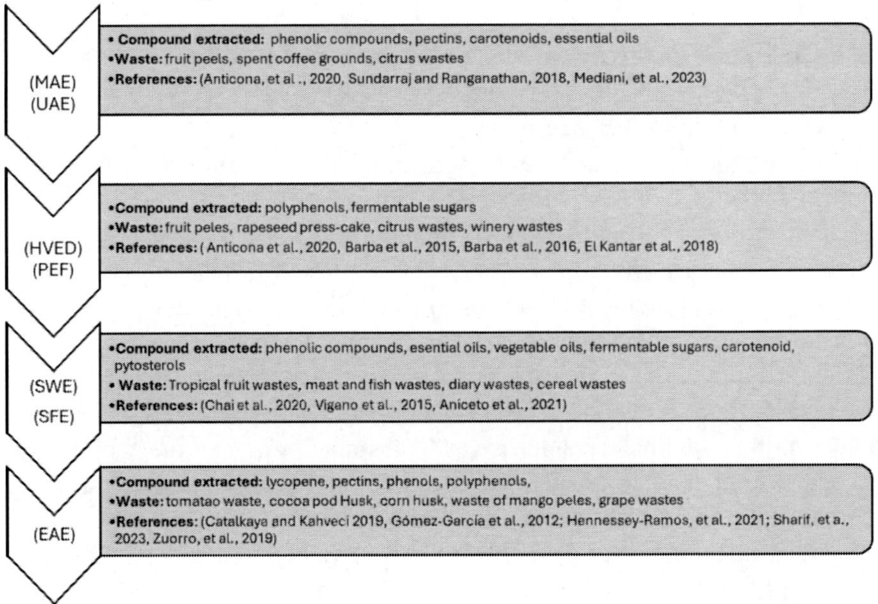

Figure 2. Non-conventional extraction methods used in the recovery of compounds of interest from agro-industrial waste.

Methods are classified as conventional, including solid-liquid extraction methods like maceration, Soxhlet extraction, hydro distillation, and steam distillation for extracting volatile compounds. These methods often use organic solvents of different polarity derived from petroleum, generating contamination residues. Additionally, they require longer extraction times and exhibit lower efficiency compared to less conventional or novel methods such as aqueous subcritical extraction (SWE), supercritical fluid extraction (SFE), microwave-assisted extraction (MAE), ultrasound-assisted extraction (UAE), pulsed electric field-assisted extraction (PFE), high-voltage electrical discharge-assisted extraction (HVED), and enzymatically assisted extraction (EAE). Most of these methods improve extraction process efficiency and allow the use of green solvents like ethanol, water, and other environmentally friendly options, leading to an increasing trend in their use for extracting components of interest from agro-industrial residues.

Numerous studies delve into this topic and can be consulted for more detailed information. Figure 2 provides examples of interesting compounds extracted from agro-industrial residues using unconventional methods.

3.4. Biological Pathway for the Valorization of Residues

3.4.1. Energetic Valorization

The energetic valorization of agro-industrial residues involves their conversion into biofuels, which can be in solid forms such as pellets and briquettes, liquid forms like bio alcohols and biodiesel, and gaseous forms such as biogas and hydrogen (Nair et al., 2022). The purpose is to reduce reliance on fossil fuels; however, its global energy contribution remains low, constrained by technological aspects. Numerous studies have explored different types of agro-industrial waste, such as sugarcane bagasse, sugar beet waste, rice straw, potato, and sweet potato waste, rich in lignocellulose, and various animal residues, as substrates for biofuel production (Yogalakshmi et al., 2022). The utilization of agricultural residues minimizes deforestation, land use competition, and pollution, yielding positive impacts on environmental, social, and economic components. These biofuels can be generated through the following biological processes:

3.4.1.1. Fermentation

Fermentation is employed to produce liquid biofuels like bioethanol and biobutanol. Microorganisms are used in this process to convert starch, cellulose, or sugar into recoverable products, such as ethanol (Cherubini, 2010). Initially, biomass is crushed, and with the application of specific enzymes based on the type of biomass, starch, cellulose, and hemicellulose are converted into 5- and 6-carbon sugars, which can be further transformed by yeasts, bacteria, or modified microorganisms into bioalcohols like ethanol. The product purification stage involves distillation.

3.4.1.2. Transesterification

Transesterification is the process in which the triglycerides in oil or fat react with an alcohol group, usually methanol, in the presence of a basic catalyst like NaOH or KOH to form glycerol and methyl esters, which constitute biodiesel. This biodiesel can be used to power diesel engines or as fuel for vehicles. It is produced from agro-industrial waste rich in fats, such as animal fats and used vegetable oils, as mentioned earlier (Azadbakht et al., 2021). Glycerol, obtained as a by-product, has pharmaceutical and cosmetic applications.

3.4.1.3. Anaerobic Digestion

Anaerobic digestion involves the degradation or decomposition of organic matter, such as agricultural and food waste, in the absence of oxygen to produce biogas, where the methane component, crucial for its calorific value, ranges between 50 and 70%. Other components include carbon dioxide, and in lesser proportions, hydrogen, hydrogen sulfide, and other gases. Parameters influencing the anaerobic digestion process include temperature, pH, total solids, retention time, volatile solids, and the Carbon/Nitrogen ratio (Cremonez et al., 2021; Martínez-Gutiérrez, 2018). Anaerobic digestion offers significant advantages for the valorization of agricultural and livestock residues; it is cost-effective, simple, and requires no external energy. However, its efficiency is contingent on the composition of the residue and decreases for materials with high lignocellulose content (Achinas and Euverink, 2016). Large-scale production of refined biogas can be directly used as a substitute for energy or for co-generation in domestic, rural, or industrial settings.

3.5. Alternative Processes to Produce High Value Bioproducts

3.5.1. Solid State Fermentation (SSF)

Agro-industrial residues can also serve as substrates for Solid State Fermentation (SSF), a conversion technique typically used in low-moisture solids. SSF offers advantages over submerged fermentation (SmF), such as low energy demand, minimal liquid effluent emission, and the possibility of use under non-sterile conditions, along with low cost. However, its industrial application is limited due to challenges in process monitoring and scaling (El-Bakry et al., 2015).

Through SSF, numerous potential products have been isolated from agricultural waste, including fruits, rice straw, wheat straw, peanut cake, etc. (Mohapatra et al., 2020; Sala et al., 2022). Among the generated products are flavors, ethanol, enzymes such as amylases, cellulases, invertases, pectinases, lipases, and proteases. Animal waste and solid waste from tanneries, cow dung, chicken feathers, and fish meal waste have been mainly used to produce proteases. These enzymes are produced by a diverse group of organisms, including filamentous fungi of the genus Trichoderma (e.g., *Trichoderma reesei*), *Aspergillus niger*, A. oryzae, bacteria like *Pseudomonas putida*, *Pseudomonas fluorescens*, *Bacillus subtilis*, and macrofungi such as *Pleurotus sajorcaju* and *Pleurotus ostreatus*, among many others (Sagar et al., 2018). All these enzymes find wide applications in the food industry, and in the pharmaceutical and cosmetic industries.

Other products such as ethanol, antibiotics, citric acid, lactic acid, and various food ingredients have also been produced through SSF using agro-industrial residues (Sheikha and Ray, 2023). Vanillin, the most important and widely used flavor in the food, cosmetic, and pharmaceutical industries, has also been produced using this fermentation method. Lindsay et al. (2022) extracted and quantified compounds such as phenylacetaldehyde, methyl benzoate, 1-octen-3-ol, and phenylethyl alcohol from the fermentation of nine different types of agro-industrial residues, achieving promising yields for this group of aromas and flavors. The use of lignocellulose-rich biomass such as sugarcane bagasse, cassava bagasse, corn cobs, wheat bran, and other cereal brans, fruit peels and shells, coffee pulp and peels, and straws and husks from various sources serve as sources of phenols through lignin hydrolysis, but also bioactive compounds such as mycophenolic acid, dicerandrol C, phenylacetates, anthraquinones, benzofurans, and alkenyl phenols that exhibit anticancer, antimicrobial, antioxidant, and antiviral activities are synthesized. Additionally, polysaccharides with antioxidant, antiproliferative, and immunomodulatory activities, as well as prebiotic effects, are produced using SSF (Verduzco-Oliva and Gutierrez-Uribe, 2020). Table 6 provides a compilation of some products obtained through SSF.

Table 6. Bioproducts derived from agro-industrial waste via SSF

Waste feedstock	Organism used in SSF	Product	Reference
Empty palm fruit bunches	*Aspergillus tubingensis* TSIP9	Fungal Lipids	(Intasit, et al., 2023)
Red swamp crab (*Procambarus clarkii*) shell waste	*Pleurotus ostreatus* CCMJ2570	Mushrooms with high protein content and Bioethanol	(Xu, et al., 2022)
Corn stalks	*Auricularia aurícula* H310	Lignocellulolytic enzymes	(Lu, et al., 2022)
Wheat straw	*Schizochytrium sp.*	Erythritol	(Liu, et al., 2021)
Rice husks, bagasse, and corn cob	*Aspergillus niger*	α-Amylase and β-Glucosidase	(Aliyah, et al., 2017)
Wheat bran	*Aspergillus niger* ATCC 13497	Feruloyl esterase	(Gopalan, et al., 2016)
Babassu residue	Exo and endoamylases, proteases, xylanases and cellulases produced by Aspergillus awamori IOC-3914	Glucose, xylose, and free amino nitrogen.	(López, et al., 2013)
Barley and Rice	*Rhizopus microsporus* var. oligosporus and *Aspergillus oryzae*	Tempeh and koji with high quality protein, increased lysine content.	(Zwinkels, Wolkers and Smid 2023)
Rice bran, fish meal and soy.	*Serratia marcescens* NCHU05	Prodigiosin	(Kuo and Li 2023)
Chicken Feathers	Recombinant *Bacillus subtilis* cells	Soluble proteins and alkaline serine protease.	(Salamony, et al., 2024)
Chicken Feathers	*Streptomyces sp.* SCUT-3	Soluble peptides and free amino acids	(Lu, et al., 2024)
Rice bran	*Clostridium tyrobutyricum* DSM 2637	Butyric acid	(Akhtar, et al., 2018)
Coffee pulp	*Saccharomyces cerevisiae*	Chlorogenic acid	(Santos, et al., 2019)
Soy Cake	*Rhizopus arrhizus* NRRL 2582	Fumaric Acid	(Papadaki, et al., 2018)

3.5.2. Bio-Based Plastics and Packaging

Over the last two decades, plastic pollution has been a topic of discussion globally, influenced by the variability of laws regulating single-use plastics and the petroleum industry's derivatives crisis. This has led to a need for alternative materials to reduce pollution and mitigate the demonstrated effects on climate change (Nandakumar, Chuah, and Sudesh, 2021). Bioplastics, defined as biological, biodegradable, or not, polymers that arise as a solution to the plastic pollution issue, present advantages such as biodegradability, making them one of the most viable options for overcoming conventional plastic pollution, although their biodegradation is still under study. Bioplastics were discovered approximately a century ago, with one of the first being microbial Polyhydroxyalkanoates (PHA), produced using apple waste, molasses, treacle, whey, olive waste, and ground coffee waste, among others (Liu et al., 2021; Razzaq et al., 2022). Currently, many bioplastics are available in the market, including polylactic acid (PLA), poly (butylene adipate-co-terephthalate) (PBAT), Mirel, and Bio-PET, thermoplastic starches, and chitosan-based bioplastics (Chan et al., 2021; Nandakumar, Chuah, and Sudesh, 2021).

Natural fibers extracted from agro-industrial waste, as shown in the table...), can also be used for the generation of biodegradable bioplastics, bio-packages, and biomaterials with biomedical applications due to their non-toxic nature. For example, derivatives of starch, and mixtures with polycaprolactone (PCL) and polybutylene succinate (PBS), have been explored. However, gradual replacement of conventional plastics has not been feasible due to production costs, posing a challenge that has led recent efforts to focus on standardization and the development of new methodologies for their production (Nandakumar, Chuah, and Sudesh, 2021).

3.5.3. Food Biomass Production

The valorization of waste for food biomass production has emerged as an innovative and sustainable approach in the context of global food security. Various strategies have been developed to harness organic waste, recovering essential nutrients, thereby contributing to closing biogeochemical cycles and alleviating pressure on natural resources. One of the most promising approaches involves cultivating edible fungi such as *Pleurotus ostreatus*, *Lentinus crinitus* (L.) Fr, among other species, on organic residues like sawdust, sugarcane bagasse, coffee pulp, and rice husk (Akter et al., 2022; Dávila et al., 2020, Davila et al., 2022). These fungi not only transform waste into biomass rich in proteins and other nutrients but also the final product or spent substrate can be used in animal feed, for recovering enzymes, bioactive compounds, or organic fertilizers (Rinker, 2017; Owaid et al., 2017). Additionally, protein production from insects, such as the black soldier fly and the red Californian earthworm, has gained interest as an alternative and sustainable source of animal feed through the conversion of various agro-industrial residues. These organisms can convert organic waste into high-quality proteins, facilitating their conversion into organic fertilizer, providing an efficient solution for waste management (Amrul et al., 2022; Muin et al., 2023; Musyoka et al., 2019; Musyoka et al., 2020).

Despite the evident advantages of these strategies, it is crucial to address associated challenges such as standardizing production processes, ensuring food safety, and gaining social acceptance for these new products. Therefore, continued research is necessary to optimize these practices, maximizing their efficiency and scalability.

3.5.4. Biofertilizer Production

Sustainable management of agro-industrial waste has emerged as a crucial strategy to mitigate environmental impacts and enhance efficiency in the agricultural sector. Transforming these residues into fertilizers represents a promising alternative, utilizing their nutritional content to enrich the soil and improve agricultural productivity. Among the most notable advantages of this practice is the reduction of environmental pollution associated with improper waste disposal, as well as contributing to the circular economy by closing the nutrient cycle.

In this regard, and aiming to maximize added value, fertilizers can be considered in the final stage of utilization, after recovering energy and material value from an agro-industrial residue. Various possibilities arise; for example, starting with the recovery of bioactive compounds, followed by an energy utilization step like biogas production, and finally using the digestate as fertilizer. Alternatively, a nutrient recovery can be initiated through food biomass production, either by using the residue for cultivating edible fungi or producing insect protein through the use of the black soldier fly. The commercial potential of Black Soldier Fly larvae and fertilizers is relatively new but has garnered significant interest in recent years within the United Nations (UN) and the European Union, as profit margins increase by obtaining protein from BSF and fertilizer (Beesigamukama et al., 2021).

Anaerobic digestion seems to be a better process than composting for converting both agricultural and livestock residues into fertilizers, as it avoids nitrogen losses through volatilization that occur during composting. However, information on this topic is still limited, necessitating further studies to understand the variability according to the type of raw material used (Zhang et al., 2021). The use of fertilizers derived from agro-industrial waste can decrease reliance on synthetic chemical fertilizers, with ensuing benefits for soil health and the quality of harvested products. Figure 2 shows some examples of the use of waste to produce fertilizers and their use as food biomass.

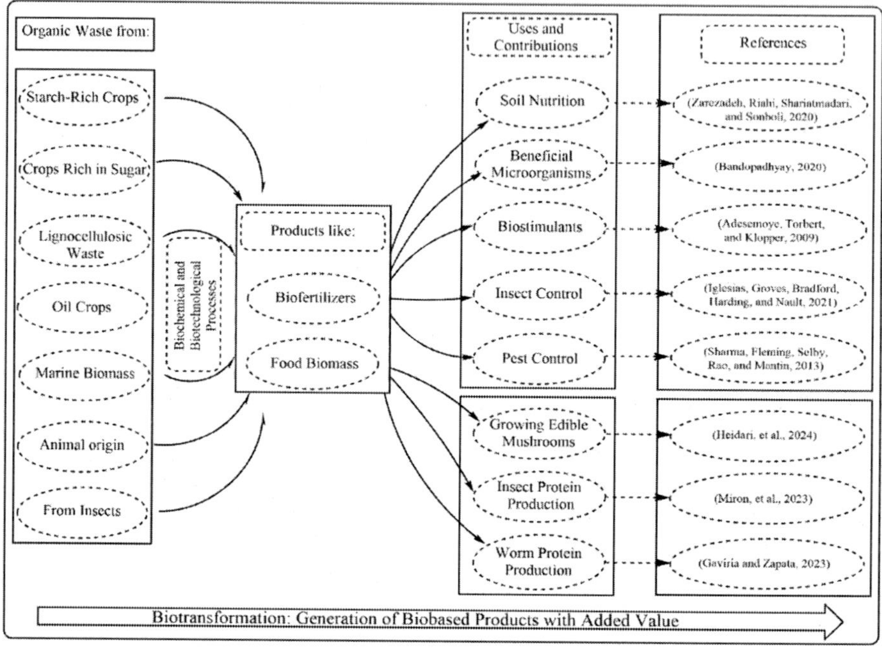

Figure 3. Examples of the use of waste to produce fertilizers and their use as food biomass.

However, it is essential to address inherent disadvantages. The variability in the composition of agro-industrial residues can result in the inconsistency of obtained fertilizers, affecting their effectiveness and specific application. Additionally, some residues may contain contaminants such as heavy metals that can concentrate in the final product, posing challenges in terms of safety and soil quality. Implementing rigorous treatment and monitoring processes is crucial to ensure the safety of fertilizers derived from agro-industrial waste.

Conclusion

The management of agro-industrial waste represents one of the primary challenges faced by contemporary societies. This is attributed not only to the environmental impact associated with the pollution they generate but also to the pressure exerted on natural resources. This pressure encompasses crucial aspects such as competition for land, water, and energy, considered from the perspective of sustainable development.

Nevertheless, various alternatives demonstrate how it is possible to transform and valorize these residues into a wide range of products. This transformation not only benefits the environment by reducing pollution but also promotes the development of new social, economic, and environmental trends. Additionally, it helps curb the impact associated with the growing demand for food production and transformation, as well as the dependence on fossil fuels.

In this context, the current chapter explores some of the alternatives for the development of bio-sustainable products through the proper management of agro-industrial waste. It focuses on the use of biotechnological strategies, which can be leveraged at different levels of scale, from initiatives in rural economies to industrial-level implementations.

Disclaimer

Nothing to declare.

References

Achinas, S., and Euverink, G. J. W. (2016). Theoretical analysis of biogas potential prediction from agricultural waste. *Resource-Efficient Technologies,* 2(3), 143-147.

Akhtar, T., Hashmi, A., Tayyab, M., Anjum, A., Saeed, S. and Ali, S. (2018). Bioconversion of Agricultural Waste to Butyric Acid Through Solid State Fermentation by *Clostridium tyrobutyricum*. *Waste and Biomass Valorization.*

Akter, M., Halawani, R. F., Aloufi, F. A., Taleb, M. A., Akter, S., and Mahmood, S. (2022). Utilization of agro-industrial wastes for the production of quality oyster mushrooms. *Sustainability,* 14(2), 994.

Aliyah, A., Alamsyah, G., Ramadhani, R. and Hermansyah, H. (2017). Production of α-Amylase and β-Glucosidase from *Aspergillus niger* by solid state fermentation method on biomass waste substrates from rice husk, bagasse and corn cob. *Energy Procedia,* 418-423.

Amrul, N. F., Kabir Ahmad, I., Ahmad Basri, N. E., Suja, F., Abdul Jalil, N. A., and Azman, N. A. (2022). A review of organic waste treatment using black soldier fly (*Hermetia illucens*). *Sustainability,* 14(8), 4565.

Aniceto, J. P., Rodrigues, V. H., Portugal, I., and Silva, C. M. (2021). Valorization of tomato residues by supercritical fluid extraction. *Processes,* 10(1), 28.

Anticona, M., Blesa, J., Frigola, A., and Esteve, M. J. (2020). High biological value compounds extraction from citrus waste with non-conventional methods. *Foods,* 9(6), 811.

Awasthi, M, et al. (2022). Biotechnological strategies for bio-transforming biosolid into resources toward circular bioeconomy: A review. *Renewable and Sustainable Energy Reviews,* 2022: 10.1016/j.rser.2021.111987.

Azadbakht, M., Safieddin Ardebili, S. M., and Rahmani, M. (2021). A study on biodiesel production using agricultural wastes and animal fats. *Biomass conversion and biorefinery,* 1-7.

Barba, F. J., Boussetta, N., and Vorobiev, E. (2015). Emerging technologies for the recovery of isothiocyanates, protein and phenolic compounds from rapeseed and rapeseed press-cake: Effect of high voltage electrical discharges. *Innovative Food Science & Emerging Technologies,* 31, 67-72.

Barba, F. J., Zhu, Z., Koubaa, M., Sant'Ana, A. S., and Orlien, V. (2016). Green alternative methods for the extraction of antioxidant bioactive compounds from winery wastes and by-products: A review. *Trends in Food Science & Technology,* 49, 96-109.

Bas-Bellver, C., Barrera, C., Betoret, N., & segui, L. (2020). Turning Agri-Food Cooperative Vegetable Residues into Functional Powdered Ingredients for the Food Industry. *Sustainability.*

Beesigamukama, D., Mochoge, B., Korir, N. K., Fiaboe, K. K., Nakimbugwe, D., Khamis, F. M., ... and Tanga, C. M. (2021). Low-cost technology for recycling agro-industrial waste into nutrient-rich organic fertilizer using black soldier fly. *Waste Management,* 119, 183-194.

Capanoglu, E., Nemli, E. and Tomas-Barberan, F. (2022). Novel Approaches in the Valorization of Agricultural Wastes and Their Applications. *J Agric Food Chem,* 2022: 6787-6804.

Carretero, A. (2018). Impactos sociales, económicos y medioambientales derivados de la pérdida y el desperdicio de alimentos. *Przegląd Prawa Rolnego,* 2018: 127-139.

Catalkaya, G., & Kahveci, D. (2019). Optimization of enzyme assisted extraction of lycopene from industrial tomato waste. *Separation and Purification Technology, 219*, 55-63.

CEC. *Acerca de - Por qué y cómo cuantificar la PDA.* s.f. http://www.cec. org/flwm/es/checklist-item-es/seleccion-de-indicadores-clave-de-

desempeno-y-determinacion-de-efectos-de-la-pda/ (último acceso: 25 de ENERO de 2024).

Cerino-Córdova, F. J., Dávila-Guzmán, N. E., León, A. M. G., Salazar-Rabago, J. J., and Soto-Regalado, E. (2020). Revalorization of coffee waste. *Coffee-production and research,* 1-26.

Chai, Y. H., Yusup, S., Kadir, W. N. A., Wong, C. Y., Rosli, S. S., Ruslan, M. S. H., ... & Yiin, C. L. (2020). Valorization of tropical biomass waste by supercritical fluid extraction technology. *Sustainability,* 13(1), 233.

Chan, J. X., Wong, J. F., Hassan, A., and Zakaria, Z. (2021). Bioplastics from agricultural waste. In *Biopolymers and biocomposites from agro-waste for packaging applications* (pp. 141-169). Woodhead Publishing.

Chen, W., et al., (2014) Hydrothermal liquefaction of mixed-culture algal biomass from wastewater treatment system into bio-crude oil. *Bioresource Technology,* 2014: 130-139.

Cherubini, F. (2010). The biorefinery concept: Using biomass instead of oil for producing energy and chemicals. *Energy Conversion and Management,* 1412-1421.

Cho, E., Phi, L., Song, Y., Gyo, Y., & Bae, H. (2020). Bioconversion of biomass waste into high value chemicals. *Bioresource Technology.*

Cremonez, P., Teleken, J., Weiser, T., & Alves, H. (2021). Two-Stage anaerobic digestion in agroindustrial waste treatment: A review. *Journal of Environmental Management.*

Dávila, G. L. R., Murillo, A. W., Zambrano, F. C. J., Suárez, M. H., and Méndez, A. J. J. (2020). *Evaluation of nutritional values of wild mushrooms and spent substrate of Lentinus crinitus (L.) Fr. Heliyon* 6: e03502.

Dávila, G. L. R., Zambrano, F. C., Torres, A. O., Pérez, J. F. B., and Murillo, A., W. (2022). Integral use of rice husks for bioconversion with white-rot fungi. *Biomass Conversion and Biorefinery,* 1-11.

Debalina, B., Rajasekhar, B., and Vinu, R. (2017). Production of carbon nanostructures in biochar, bio-oil and gases from bagasse via microwave assisted pyrolysis using Fe and Co as susceptors. *Journal of Analytical and Applied Pyrolysis,* 2017: 310-318.

Ebrahumi, P., Mihaylova, D. and Lante, A. (2022). Comparison of green technologies for valorizing sugar beet (*Beta vulgaris* L.) leaves. *Food Science and Applied Biotechnology*, 2022: 119-130.

El Kantar, S., Boussetta, N., Rajha, H. N., Maroun, R. G., Louka, N., & Vorobiev, E. (2018). High voltage electrical discharges combined with enzymatic hydrolysis for extraction of polyphenols and fermentable sugars from orange peels. *Food Research International,* 107, 755-762.

El-Bakry, M., Abraham, J., Cerda, A., Barrena, R., Ponsá, S., Gea, T., and Sánchez, A. (2015). From wastes to high value-added products: novel aspects of SSF in the production of enzymes. *Critical Reviews in Environmental Science and Technology,* 45(18), 1999-2042.

FAO, FIDA, OMS, PMA, and UNICEF. Versión resumida de El estado de la seguridad alimentaria y la nutrición en el mundo 2023. *Urbanización, transformación de los sistemas agroalimentarios y dietas saludables a lo largo del continuo rural-urbano.* Roma: FAO, 2023.

FAO. (2011). *El estado de la inseguridad alimentaria en el mundo.* Roma: FAO.

FAO. (2019). *El estado mundial de la agricultura y la alimentación. Progresos en la lucha contra la pérdida y el desperdicio de alimentos.* Roma: FAO.

FAO. *Organización de las Naciones Unidas.* 2023. https://www.un.org/es/global-issues/food (último acceso: 15 de enero de 2024).

Gómez-García, R., Martínez-Ávila, G. C., & Aguilar, C. N. (2012). Enzyme-assisted extraction of antioxidative phenolics from grape (Vitis vinifera L.) residues. *3 Biotech,* 2, 297-300.

Gopalan, N., Nampoothiri, K., Szakacs, G., Parameswaran, B. and Pandey, A. (2016). Solid-state fermentation for the production of biomass valorizing feruloyl esterase. *Biocatalysis and Agricultural Biotechnology,* 2016: 7-13.

Gustavsson, J, C Cederberg, and U Sonesson. *Pérdidas y desperdicio de alimentos en el mundo: Alcance, causas y prevención.* FAO, 2011.

Hegnsholt, E., Unnikrishnan, S., Pollmann-Larsen, M., Askelsdottir, B., and Gerard, M. (2018). *Tackling the 1.6-Billion-Ton Food Loss and Waste Crisis.* Boston Consulting Group. Retrieved from https://www.bcg.com/publications/2018/tackling-1.6-billion-ton-food-loss-and-waste-crisis.

Hennessey-Ramos, L., Murillo-Arango, W., Vasco-Correa, J., & Paz Astudillo, I. C. (2021). Enzymatic extraction and characterization of pectin from cocoa pod husks (Theobroma cacao L.) using Celluclast® 1.5 L. *Molecules,* 26(5), 1473.

Hernández-Carmona, F., Morales-Matos, Y., Lambis-Miranda, H., and Pasqualino, J. (2017). Starch extraction potential from plantain peel wastes. *Journal of environmental chemical engineering,* 5(5), 4980-4985.

Igbokwe, P. K., Idogwu, C. N., and Nwabanne, J. T. (2016). Enzymatic hydrolysis and fermentation of plantain peels: optimization and kinetic studies. *Advances in Chemical Engineering and Science,* 6(2), 216-235.

Intasit, R., B Cherisilp, Y louhasakul, and N Thongchul. (2023). Enhanced biovalorization of palm biomass wastes as biodiesel feedstocks through integrated solid-state and submerged fermentations by fungal co-cultures. *Bioresource Technology.*

Khan, F., Ali, y Y. (2022). A facilitating framework for a developing country to adopt smart waste management in the context of circular economy. Environmental Science and Pollution Research, 2022: 26336 - 26351.

Kolesárová, N., Huntnan, M., Bodik, I., & Spalková, V. (2011). Utilization of Biodiesel By-Products for Biogas Production. *J Biomed Biotechnol.*

Kuo, Y, y S Li. (2023). Solid-state fermentation of food waste by Serratia marcescens NCHU05 for prodigiosin production. *Journal of the Taiwan Institute of Chemical Engineers.*

Lindsay, M. A., Granucci, N., Greenwood, D. R., and Villas-Boas, S. G. (2022). Identification of new natural sources of flavour and aroma metabolites from solid-state fermentation of agro-industrial by-products. *Metabolites,* 12(2), 157.

Liu, H., Kumar, V., Jia, L., Sarsaiya, S., Kumar, D., Juneja, A., ... and Awasthi, M. K. (2021). Biopolymer poly-hydroxyalkanoates (PHA) production from apple industrial waste residues: A review. *Chemosphere,* 284, 131427.

Liu, X., et al., (2021). Enhancing erythritol production by wheat straw biochar-incorporated solid-state fermentation of agricultural wastes using defatted *Schizochytrium sp.* biomass as supplementary feedstock. *Industrial Crops and Products.*

López, J., Lázaro, C., dos Reis, L., Guuimaraes, D. and Machado, A. (2013). Characterization of multienzyme solutions produced by solid-state fermentation of babassu cake, for use in cold hydrolysis of raw biomass. *Biochemical Engineering Journal,* 2013: 231-239.

Lu, W., et al., (2024). Sustainable valorizing high-protein feather waste utilization through solid-state fermentation by keratinase-enhanced *Streptomyces sp.* SCUT-3 using a novel promoter. *Waste Management,* 2024: 528-538.

Lu, X., Li, F., Zhou, X., Hu, J. and Liu, P. (2022). Biomass, lignocellulolytic enzyme production and lignocellulose degradation patterns by *Auricularia auricula* during solid state fermentation of corn stalk residues under different pretreatments. *Food Chemistry.*

Martínez-Gutiérrez, E. (2018). Biogas production from different lignocellulosic biomass sources: advances and perspectives. *Biotech,* 8(5), 233.

Mediani, A., Kamal, N., Lee, S. Y., Abas, F., & Farag, M. A. (2023). Green extraction methods for isolation of bioactive substances from coffee seed and spent. *Separation & Purification Reviews,* 52(1), 24-42.

Mohapatra, S., Pattnaik, S., Maity, S., Mohapatra, S., Sharma, S., Akhtar, J., . . . Varma, A. (2020). Comparative analysis of PHAs production by Bacillus megaterium OUAT 016 under submerged and solid-state fermentation. *Saudi Journal of Biological Sciences,* 1242-1250.

Muin, H., Alias, Z., Nor, A. M., and Taufek, N. M. (2023). Bioconversion of Desiccated Coconut and Soybean Curd Residues for Enhanced Black Soldier Fly Larvae Biomass as a Circular Bioeconomy Approach. *Waste and Biomass Valorization,* 14(1), 249-260.

Musyoka, S. N., Liti, D. M., Ogello, E., and Waidbacher, H. (2019). Utilization of the earthworm, *Eisenia fetida* (Savigny, 1826) as an alternative protein source in fish feeds processing: A review. *Aquaculture Research,* 50(9), 2301-2315.

Musyoka, S. N., Liti, D., Ogello, E. O., Meulenbroek, P., and Waidbacher, H. (2020). Earthworm, *Eisenia fetida*, bedding meal as potential cheap fishmeal replacement ingredient for semi-intensive farming of Nile Tilapia, Oreochromis niloticus. *Aquaculture Research,* 51(6), 2359-2368.

Nair, L. G., Agrawal, K., and Verma, P. (2022). An overview of sustainable approaches for bioenergy production from agro-industrial wastes. *Energy Nexus,* 6, 100086.

Nandakumar, A., Chuah, J., & Sudesh, K. (2021). *Bioplastics: A boon or bane? Renewable and Sustainable Energy Reviews.*

ODELA. (2019). *Polisemias de la alimentacion Salud, desperdicio, hambre y patrimonio.* Barcelona: Universitar De Barcelona Editions.

Owaid, M. N., Abed, I. A., and Al-Saeedi, S. S. S. (2017). Applicable properties of the bio-fertilizer spent mushroom substrate in organic systems as a byproduct from the cultivation of *Pleurotus spp. Information Processing in Agriculture,* 4(1), 78-82.

Papadaki, A., y otros. (2018). Fumaric acid production using renewable resources from biodiese and cane sugar production processes. *Environmental Science and Pollution Research.*

Pappas, V., et al. (2021). Pressurized Liquid Extraction of Polyphenols and Anthocyanins from Saffron Processing Waste with Aqueous Organic Acid Solutions: Comparison with Stirred-Tank and Ultrasound-Assisted Techniques. *Sustainability.*

Pérez, A., D. Leyva, and F. Gómez. (2018). Desafíos y propuestas para lograr la seguridad alimentaria. *Revista Mexicana de Ciencias Agrícolas,* 2018.

Pinto, J., Boavida-Dias, R., Matos, H. A., and Azevedo, J. (2022). Analysis of the food loss and waste valorisation of animal by-products from the retail sector. *Sustainability,* 14(5), 2830.

Razzaq, S., Shahid, S., Farooq, R., Noreen, S., Perveen, S., and Bilal, M. (2022). Sustainable bioconversion of agricultural waste substrates into poly (3-hydroxyhexanoate) (mcl-PHA) by Cupriavidus necator DSM 428. *Biomass Conversion and Biorefinery, 1*-11.

Rinker, D. L. (2017). Spent mushroom substrate uses. Edible and medicinal mushrooms: technology and applications, 427-454.

Sagar, N. A., Pareek, S., Sharma, S., Yahia, E. M., and Lobo, M. G. (2018). Fruit and vegetable waste: Bioactive compounds, their extraction, and possible utilization. *Comprehensive reviews in food science and food safety,* 17(3), 512-531.

Sala, A., Echegaray, T., Palomas, G., Boggione, M., Tubio, G., Barrera, R., & Artola, A. (2022). Insights on fungal solid-state fermentation for waste valorization: Conidia and chitinase production in different reactor configurations. *Sustainable Chemistry and Pharmacy.*

Salamony, D., Eldin, M., Zaghloul, T. and Moustafa, H. (2024). Valorization of chicken feather waste using recombinant bacillus subtilis cells by solid-state fermentation for soluble proteins and serine alkaline protease production. *Bioresource Technology.*

Santos, J., et al. (2019). Solid-state fermentation as a sustainable method for coffee pulp treatment and production of an extract rich in chlorogenic acids. *Food and Bioproducts Processing,* 2019: 175-184.

Sato, M. F., Vieira, R. G., Zardo, D. M., Falcão, L. D., Nogueira, A., and Wosiacki, G. (2010). Apple pomace from eleven cultivars: an approach to identify sources of bioactive compounds. Acta Scientiarum. *Agronomy,* 32, 29-35.

Sathish, R., Sasikumar, R. and Dhilipkumar, T. (2024). Exploiting agro-waste for cleaner production: A review focusing on biofuel generation, bio-composite production, and environmental considerations. *Journal of Cleaner Production,* 2024: 10.1016/j.jclepro.2023.140536.

Shahabuddin, M., Krishna, B., Bhaskar, T. and Perkins, G. (2020). Advances in the thermo-chemical production of hydrogen from biomass and residual wastes: Summary of recent techno-economic analyses. *Bioresource Technology.*

Shahid, M., Batool, A., Kashif, A., Nawaz, M., Aslam, M., Iqbal, N., and Choi, Y. (2021). Biofuels and biorefineries: Development, application and future perspectives emphasizing the environmental and economic aspects. *Journal of Environmental Management,* 113268.

Sharif, T., Bhatti, H. N., Bull, I. D., & Bilal, M. (2023). Recovery of high-value bioactive phytochemicals from agro-waste of mango (Mangifera indica L.) using enzyme-assisted ultrasound pretreated extraction. *Biomass Conversion and Biorefinery,* 13(8), 6591-6599.

Sharma, H., Fleming, C., Selby, C., Rao, J., & Mantin, T. (2013). Plant biostimulants: a review on the processing of macroalgae and use of extracts for crop management to reduce abiotic and biotic stresses. *Journal of Applied Phycology,* 465-490.

Sheikha, A. F., and Ray, R. C. (2023). Bioprocessing of horticultural wastes by solid-state fermentation into value-added/innovative bioproducts: A review. *Food Reviews International,* 39(6), 3009-3065.

Sivakumar, D., Srikanth, P., Ramteke, P. W., and Nouri, J. (2022). Agricultural waste management generated by agro-based industries using biotechnology tools. *Global Journal of Environmental Science and Management,* 8(2), 281-296.

Spinei, M., and Oroian, M. (2021). The potential of grape pomace varieties as a dietary source of pectic substances. *Foods,* 10(4), 867.

Statista. (n.d.). *Production of meat worldwide from 2016 to 2023, by type* (in million metric tons). Retrieved from https://www.statista.com/statistics/237632/production-of-meat-worldwide-since-1990/.

Sundarraj, A. A., and Ranganathan, T. V. (2018). Comprehensive review on ultrasound and microwave extraction of pectin from agro-industrial wastes. *Drug Invention Today,* 10.

UNEP, Informe sobre el índice de desperdicios de alimentos 2021. *Nairobi: Programa de las Naciones Unidas para el Medio Ambiente,* 2021.

Verduzco-Oliva, R., and Gutierrez-Uribe, J. A. (2020). Beyond enzyme production: Solid state fermentation (SSF) as an alternative approach to produce antioxidant polysaccharides. *Sustainability,* 12(2), 495.

Vigano, J., da Fonseca Machado, A. P., & Martinez, J. (2015). Sub-and supercritical fluid technology applied to food waste processing. *The Journal of Supercritical Fluids,* 96, 272-286.

Xu, J., Kim, J. and Choi, Y. (2019). Simultaneous roasting and extraction of green coffee beans by pressurized liquid extraction. *Food Chemistry,* 2019: 261-268.

Xu, S., Gao, M., Peng, Z., Sui, K., Li, Y. and Li, C. (2022). Upcycling from chitin-waste biomass into bioethanol and mushroom via solid-state fermentation with Pleurotus ostreatus. *Fuel.*

Yogalakshmi, K., Devi, P., Sivashanmugam, P., Kavitha, S., Kannah, Y., Varjani, S., . . . Banu, R. (2022). *Lignocellulosic biomass-based pyrolysis: A comprehensive review. Chemosphere.*

Zhang, X., Lopes, I. M., Ni, J. Q., Yuan, Y., Huang, C. H., Smith, D. R., ... and Wu, S. (2021). Long-term performance of three mesophilic anaerobic digesters to convert animal and agro-industrial wastes into organic fertilizer. *Journal of cleaner production,* 307, 127271.

Zuorro, A., Lavecchia, R., González-Delgado, Á. D., García-Martinez, J. B., & L'Abbate, P. (2019). Optimization of enzyme-assisted extraction of flavonoids from corn husks. *Processes,* 7(11), 804.

Zwinkels, J., Wolkers, J. and Smid, E. (2023). Solid-state fungal fermentation transforms low-quality plant-based foods into products with improved protein quality. *LWT.*

Chapter 2

Application of Mushroom By-Products in Livestock Diets

Agori Karageorgou, MSc
Maria Papageorgiou, MSc
Michael Goliomytis, PhD
and Panagiotis Simitzis*, PhD
Laboratory of Animal Breeding and Husbandry, Department of Animal Science, School of Animal Biosciences, Agricultural University of Athens, Athens, Greece

Abstract

The continuing increase of human population has resulted in a continuous demand for great quantities of agricultural and livestock products, and animal feed for livestock species. Consequently, the volume of agro-industrial waste is increasing. This fact has a negative impact on the environment and has become a concern that attracts the attention of both society and the scientific community. Adding value to agro-industrial wastes is not only environmentally, but also economically sustainable since it contributes to the circular economy. For these reasons, there is a growing interest on research for the development of innovative and sustainable methods of the exploitation and valorization of agro-industrial wastes, especially solid ones, including ways to successfully utilize them for animal nutrition. Mushrooms have particularly gained attention, due to their ability to turn waste into food of high quality with a beneficial and health promoting nutritional profile. During the industrial processing of mushrooms, a great number of by-products or wastes are produced.

* Corresponding Author's Email: pansimitzis@aua.gr.

In: Title Agro-Industrial Wastes
Editor: Peter Clements
ISBN: 979-8-89113-688-5

There is growing research that focuses on the management of these by-products and their environmental impact when inadequately disposed of. These by-products could be incorporated into livestock diets to enhance the nutritional profile of the derived products and reduce the environmental impact of their disposal. As indicated, they also possess antibacterial, antioxidant and immunomodulatory properties that could boost animal health status. The main purpose of this review is therefore to highlight the relative composition of mushroom by-products and their nutritional value. Spent mushroom substrates could be especially used for the feeding of ruminants, due to their high lignin content, but also in poultry and pigs after application of biological methods such as fermentation that improve nutritional value by enhancing crude protein, fat and ash content and lowering the amount of crude fibre.

Keywords: mushrooms, spent substrate, livestock, products' quality; sustainability, circular economy, valorization

Introduction

The increasing global population necessitates improved management of agricultural and animal production systems to meet food demand. Over 820 million people experienced hunger in 2018, and 2 billion face food insecurity (FAO, 2019). As for other commodities, production of mushrooms and truffles has therefore been increased worldwide by 65% during the previous decade; it reached around 48.34 million metric tons (MT) in 2022 from 31.78 million MT in 2012 (FAOSTAT, 2024). China with 45.43 million MT was the dominant mushroom producer country in 2022, while in the EU, Poland and Netherlands were the top two producers with 257 and 235 thousand MT, respectively (FAOSTAT, 2024).

The most widely grown mushrooms globally are *Pleurotus* species, *Lentinula edodes (Berk.) Pegler* and *Agaricus bisporus (J.E. Lange) Imbach* (Kamarudzaman et al., 2015). *Pleurotus* species, which are grown especially in Asia, America, and Europe, are rich in protein, polysaccharides, fiber, minerals, and vitamins and display antiviral, antitumor, antibacterial, hypocholesterolemic, immunomodulatory and activities (Cohen et al., 2002). As saprophytes, they extract nutrients from their medium, including nitrogen, carbon, minerals, vitamins, phosphorus, magnesium, potassium, and trace elements like selenium, iron, copper, zinc, manganese, and molybdenum (Zakil et al., 2022). Especially, *Pleurotus ostreatus* has been found to possess anti-inflammatory (Taofiq et al., 2015; Masri et al., 2017), antioxidant (Stefan et al., 2015; Masri et al., 2017), anti-viral, anti-bacterial, anti-nociceptive, anti-hyperglycemic, anti-platelet aggregating, anti-proliferative, anti-atherosclerotic, anti-thrombotic, anti-osteoporotic (Masri et al., 2017), immunomodulatory and anti-tumor properties (Facchini et al., 2014; Masri et al., 2017). *Pleurotus eryngii,*, known as the "King Oyster mushroom" also contains secondary metabolites like phenolic compounds and polysaccharides that display antioxidative (Lee et al., 2012; Li and Shah, 2013; Zhang et al., 2020) and anti-inflammatory (Chuang et al., 2021) activities. As *Pleurotus* species, *Lentinula edodes* also possesses anti-tumor, immune-stimulatory, hypocholesterolemic, antimicrobial (Hatvani et al., 2002; Manzi & Pizzoferatto, 2000; Mau et al., 2001; Yang et al., 2002; Baba et al., 2015), antioxidant (Hatvani et al., 2002; Manzi & Pizzoferatto, 2000; Mau et al., 2001; Yang et al., 2002; Kitzberger et al., 2007; Reis et al., 2012), antibacterial (Parola et al., 2017) and antifungal (Kitzberger et al., 2007) properties. Finally, *Agaricus bisporus* also exhibits anticancer, antibacterial, immunomodulatory, anti-inflammatory, and antioxidant activities (Stojkovic et al., 2014). In general, the moisture, ash, protein, fat and carbohydrate content of the most common mushrooms, namely *Agaricus bisporus, Pleurotus ostreatus, P. eryngii, Lentinula edodes* and *Flammulina velutipes* is within 79.78-91.64, 0.62-1.36, 0.47-1.29, 0.14-0.35, 5.98-17.62%, respectively. It is very important to take into consideration that the composition greatly varies depending on location, type of mushroom and other factors (Reis et al., 2012).

The fruiting body of the mushrooms is the commercial final product. However, various by-products remain from the cultivation process in the fermented substrate. The main by-products are fruiting bodies that do not comply with the commercial standards, mycelium, stem and spent mushroom substrate (SMS), which is the main byproduct in volume (Figure 1) (Antunes et al., 2020). Nan et al. (2015) determined the nutrient content of mushroom residues, i.e., stems and fruiting bodies that do not meet the commercial standards, of 10 different mushroom species after hot water extraction, and found that they contained 30–50% carbohydrates and 10–25% crude protein. Moreover, approximately 5 kg of SMS are produced for each kilogram of fresh mushroom (Zisopoulos et al., 2016). SMS is defined as the exhausted residual biomass generated by commercial mushroom industries after harvesting of mushroom fruiting bodies and contains approximately 35% dry matter, 20% organic matter and 13% ash (Gerrits, 1993). As it can be concluded, tons of agricultural waste related to mushroom cultivation are accumulated. These wastes are potential environmental pollutants if they are not appropriately handled, while their treatment is also financially demanding (Ritota and Manzi, 2019). In the past, discarding by disposal and burning methods were used that are nowadays not accepted in the context of sustainability and protection of natural resources (Mohd Hanafi et al., 2018). Efficient recycling and valorization of SMS is therefore necessary in the frame of the circular economy principles (Martin et al., 2023).

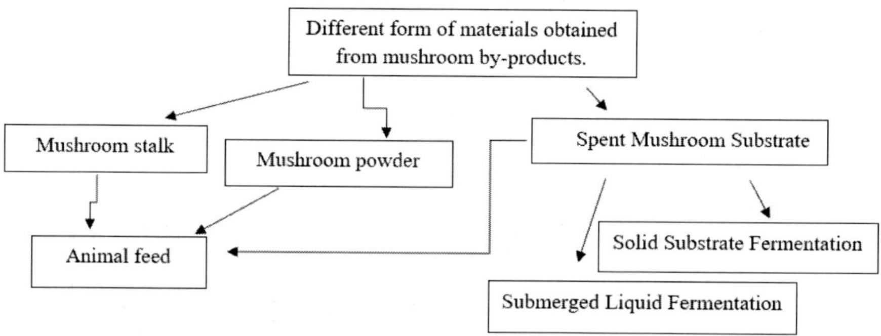

Figure 1. Different types of mushroom by-products used in livestock diets.

SMS composition and properties are influenced by the raw materials and supplements used to prepare the initial mushroom substrate. In general, it is a lignocellulosic biomass, i.e., wheat, barley or oat straw that contains several useful nutrients, such as polysaccharides, i.e., cellulose and hemicellulose, proteins, vitamins and some trace elements, namely iron (Fe), calcium (Ca), zinc (Zn) and magnesium (Mg). However, SMS is also abundant in lignin that could diminish feed digestibility, so treatments such as fermentation technique could be applied before its incorporation into animal diets to improve its nutritional value by enhancing crude protein, fat and ash content and lowering the amount of crude fibre leading to an increase of supplementation level (Abdel-Aziz et al., 2015).

The high cost of animal feed in semi-arid and arid countries poses a significant challenge to livestock production. As a result, utilizing agro-industrial by-products as alternative feed resources is a practical approach in low-input systems. This strategy aims to overcome the issue of expensive animal conventional feedstuffs and provide a cost-effective solution for livestock farmers in these regions. By the application of agro-industrial by-products, such as crop residues or agricultural waste, farmers can reduce their reliance on costly feed options and improve the sustainability of their livestock production. This approach not only helps to mitigate the financial burden on farmers but also promotes the efficient use of available resources and contributes to the overall development of a resilient livestock sector in semi-arid and arid countries (Simitzis and Deligeorgis, 2018).

Application of Mushroom By-Products in Ruminants

Cattle

Pleurotus ostreatus SMS has been evaluated as an ingredient in cattle diets as raw, just dried and milled (Baek et al., 2017) or after fermentation with bacterial inoculation (Baek et al., 2017; Kim et al., 2011). Different properties have been attributed to the fermented SMS depending on the bacterium strain that is used for the inoculation. Baek et al. (2017) used *Lactobacillus brevis* in order to decrease pH, prevent the putrefying of the crops due their high moisture content, and consequently increase the preservation of SMS and at the same time its digestibility. Nonetheless, the ruminal fermentation and nutrients' digestibility were the same in Hanno steers that were fed inoculated or no inoculated SMS. However, it was observed that the inoculation increased the degradation rate of pure protein. Baek et al. (2017) suggested that ensiled *Pleurotus ostreatus* SMS inoculated with *Lactobacillus brevis* is a cost-effective solution for cow diets and the specific bacterium strain increases the preservation and the storage time of SMS by five days, since pH was decreased at the value of 4.5. Kim et al. (2011) also tried to increase the storage time of SMS by lowering its pH. Various strains were tested and the combination of *Lactobacillus plantarum Lp1', Pediococcus acidilactici Pa193* and *L. plantarum Lp2M* provided the best results. This combination resulted in a final pH of 3.81 and a storage time of 17 days. The inoculated SMS was also tested as postweaning feed for dairy calves. The calves that were fed 10% of the inoculated SMS displayed the highest growth performance, immunoglobulin A and G, hemoglobin, hematocrit, and red blood cell concentration. Therefore, the inoculated SMS was also recommended as a feed supplement that can strengthen the immune system and improve health status and postweaning growth performance.

Another mushroom strain that has been assessed as an ingredient for cattle diets is *Pleurotus eryngii*. Kim et al. (2012) inoculated *Pleurotus eryngii* SMS with *Enterobacter spp.* and *Bacillus spp.* and evaluated it as a feed ingredient for Hanoo steers. Not only the growth performance was increased, but also the carcass traits, with the ribeye area being higher in the steers that consumed the SMS. In this study, the animals that had the highest intake of the inoculated SMS had *ad libitum* access only to SMS and rice. The voluntary intake of SMS was increased with age while the voluntary intake of rice was decreased. Rangubhet et al. (2017) examined *Flammulina velutipes* SMS mixed with crop corn and ensiled as feed for Holstein steers. The animals that were fed the ensiled feed had lower rumen protozoa population, rumen acetate and enteric methane emission. According to Rangubhet et al. (2017), although the mechanism that caused these reductions is not fully understood, it is possible that they have been caused by the bioactive substances contained in the SMS.

In addition to the use of SMS in cattle diets either milled (Baek et al. 2017) or fermented (Baek et al., 2017; Kim et al. 2011; 2012; Rangubhet et al. 2017), some studies refer to the use of SMS water extract (Liu et al., 2015a; 2015b). The procedure includes water extraction, concentration under vacuum conditions and then spray drying to powder form. Two mushroom strains have been used, *Ganoderma lucidum* (Liu et al., 2015a) and *Ganoderma balabacense* (Liu et al., 2015b), both tested in dairy cattle. In both studies the addition of the powder was 33, 67 and 100 g per cow daily. In all three groups, *Ganoderma lucidum* water extract increased immunoglobulin A and serum antioxidant capacity (Liu et al., 2015a). *Ganoderma balabacense* extract increased milk yield, milk protein, triglycerides concentration, and the somatic cell count, thus displaying anti-inflammatory properties (Liu et al., 2015b).

The literature regarding the utilization of mushroom by-products in cattle is summarized in Table 1.

Sheep

Pleurotus ostreatus has been proposed as a feed ingredient suitable for lamp feeding after drying (Aldoori et al., 2015) or composting with other ingredients (Ehtesham and Vakili, 2015). Aldoori et al. (2015) harvested the SMS by-product, dried and incorporated it into the ration of Awassi lambs, up to 20%. The animals were also fed straw ad libitum and barley, the amount of which was adapted depending on the amount of SMS so that all treatment groups would have the same fixed amount of protein.

Table 1. Application of mushroom by-products in ruminants

Species	Mushroom strain	Form	Level	Results	Reference
Holstein steers	*Flammulina velutipes*	SMS-based silage	80% of bermuda hay	Decrease in total protozoa population, rumen acetate and methane emission	Rangubhet et al., 2017
Dairy cows	*Ganoderma lucidum*	SMS hot water extract	33, 67 or 100 g/day	No effect on milk IgG, IgA, or IgM levels, but increase of serum IgA and total antioxidant capacity	Liu et al., 2015a
Dairy cows	*Ganoderma balabacense*	SMS hot water extract	33, 67 or 100 g/day	Increase of milk yield, milk protein and triglyceride levels, decrease of somatic cell count and alanine aminotransferase (100 g/day)	Liu et al., 2015b
Hanoo steers	*Pleurotus eryngii*	Microbially-fermented SMS	50% of rice straw	Increase of dry matter intake, weight gain and rib eye area	Kim et al., 2012
Hanoo steers	*Pleurotus ostreatus*	SMS inoculated with Lactobacillus brevis	15%	No effect on ruminal fermentation characteristics and total nutrients digestibility	Baek et al., 2017
Holstein calves	*Pleurotus ostreatus*	SMS fermented with Lactobacillus plantarum Lp1, Pediococcus acidilacticii Pa193, L. plantarum Lp2M	10%	Improvement of average daily gain and feed efficiency. No effect on IgG, IgA, red blood cell, hemoglobin and hematocrit	Kim et al., 2011
Mehraban male lambs	*Agaricus bisporus*	SMS compost silage treated with molasses	Ad libitum	Improvement of nutritional value	Zaboli et al., 2023
Ewes	*Flammulina velutipes*	Ensiled SMS containing apple pomace or sweet corn stover	Ad libitum	Decreased voluntary intake	Koh, 2023
Hu lambs	*Pleurotus eryngii*	Fermented SMS	15, 30 and 45%	Improvement of the production performance, meat quality and rumen bacterial community diversity and abundance	Huang et al., 2023

Table 1. (Continued).

Species	Mushroom strain	Form	Level	Results	Reference
Lambs	*Pleurotus eryngii*	Ensiled SMS	40% (full replacement of rye straw	Increase of protein metabolism and utilization, and decrease of fiber digestibility	Seok et al., 2016
Awassi lambs	*Pleurotus ostreatus*	SMS	5, 10, 15 or 20%	Decrease of hot and cold carcass weight, dressing percentage, backfat thickness, rib eye area and leg lean percentage at 15-20%	Aldoori et al., 2015
Kurdish male lambs	*Pleurotus ostreatus*	SMS	15, 25 or 35%	No effect on serum biochemical and hematological parameters. Improvement of average daily gain and final weight at 25%	Ehtesham and Vakili, 2015
Berari meat goats	*Agaricus bisporus*	SMS	5 or 10% of protein	No effect in body weight gain and nutrient digestibility. Improvement of body condition score, hemoglobin, glucose and net profit at 10%	Mandale et al., 2023
Guizhou black goats	*Flammulina velutipes*	SMS	30%	Improvement of rumen fermentation, rumen microbiota, growth performance, apparent digestibility and immunity	Long et al., 2024
Liuyang black goats	*Pleurotus ostreatus*	Fermented SMS with plant rice	35%	No effect on dry matter intake. Increase of average daily gain, serum albumin content and alkaline phosphatase activity and reduction of the feed to gain ratio	Huang et al., 2022
Alpine dairy goats	*Pleurotus sajor-caju*	Fermented SMS with rice straw	15%	No effect on rumen pH value, ammonia nitrogen level, dry matter intake and milk yield	Fan et al., 2022
Female sika deer	*Pleurotus ostreatus*	SMS	10, 20 and 30%	No effect on growth performance	Yuan et al., 2022a
Male sika deer	*Flammulina velutipes*	SMS	10%	No effect on apparent nutrient digestibility, feed intake, velvet antler production and biochemical indices	Yuan et al., 2022b
Elk	*Pleurotus eryngii*	SMS	15 or 20%	Increase of blood monocytes, hemoglobin, hematocrit, serum blood urea nitrogen, glucose and high-density lipoprotein cholesterol levels	Park et al., 2012

After slaughter, lambs that were fed SMS in a percentage higher than 15% w/w had lower body, hot and cold carcass weight, rib eye area and backfat thickness. According to the results, the authors proposed that *Pleurotus ostreatus* SMS can be used in the nutrition of lambs and serve as a cost-effective solution, however in a percentage less than 15%. Ethesham and Vakili (2015) composted a wheat straw-based mixture with *Pleurotus ostreatus* and evaluated it as feedstuff for Kurdish male lambs. It was used up to 35% in their ration and had no effect on hematocrit and hemoglobin levels. Ethesham and Vakili (2015) also observed no effect on cholesterol and total triglycerides concentration. Furthermore, Ethesham and Vakili (2015) reported that the weight gain of the lambs was decreased when the composted SMS was used in a percentage higher that 25%. According to the authors, this can be explained by the fact that the composted SMS had high ash and limited minerals' content. Thus, the nutritional value was reduced, and consequently the growth of the animals when used at a higher level than 25% in the ration.

Another SMS mushroom strain that has been studied as feed ingredient in lamb diet is that of *Pleurotus eryngii*, mainly composted (Huang et al., 2023; Seok et al., 2016). Seok et al. (2016) used SMS composted with rye straw and recycled poultry bedding. In this study the SMS was used at 45% in the compost, the highest percentage compared to all other studies. A mixture of microorganisms was used for the fermentation. The composted SMS substituted rye straw in lambs' diet. The substitution led to higher protein digestibility and utilization, higher nitrogen retention but lower dry matter and so lower fiber digestibility. Huang et al. (2023) used composted *Pleurotus eryngii* SMS as feed supplement, again up to 45%, in the diet of Hu lambs. The richness and diversity of rumen bacteria was improved, and the addition of 30% composted SMS gave the best results. Moreover, the feed conversion ratio and the meat quality were improved.

Flammulina velutipes SMS has also been evaluated as a feed ingredient of sheep diet (Koh, 2023). The author cultivated this strain on agricultural sweet corn by-products, ensiled the produced SMS and fed it to ewes. The cultivation of sweet corn resulted in higher feed intake, ruminating, and chewing time compared to ewes that were fed conventional silage, with no sweet corn. Finally, *Agaricus bisporus* SMS has also been researched as feedstuff for sheep. Zaboli et al. (2023) studied the addition of SMS compost silage after treatment with various percentages of molasses in the diet of

Mehraban lambs. The authors concluded that ensiling the compost increased the ruminal fermentation and the digestibility of nutrients.

Studies concerning the utilization of mushroom by-products in sheep are summarized in Table 1.

Goats

Fan et al. (2022) investigated the potential of improving the nutritional content of rice straw as feed for goats by fermentation with *Pleurotus sajor-caju*. The composted SMS improved the rumen digestibility and decreased the concentration of volatile fatty acids, while pH was not affected. Furthermore, dry matter intake and milk yield were increased. Huang et al. (2023) incorporated *Pleurotus ostreatus* SMS after co-fermentation with rice in the diet of Liuyang black goats. The combination of the two ingredients resulted in the highest feed intake, the lowest feed conversion ratio, and the highest serum albumin levels compared to the sole SMS or rice silage utilization in the diet. Mandale et al. (2023) used dried mushroom waste of *Agaricus bisporus* in the rations of Berari meat goats, mainly as protein source in the diet. It was observed that using the by-product as 10% w/w protein source led to improved body condition score and enhanced immune response, although the nutrient digestibility was not affected (Table 1).

Deer and Elks

Pleurotus ostreatus (Yuan et al. 2022a) and *Flammulina velutypes* (Yuan et al. 2022b) SMS have both been proposed as a mean to reduce the feed cost in deer diets. Yuan et al. (2022a) replaced the standard concentrated supplement (based on corn and soybean meal) with ensiled *Pleurotus ostreatus* SMS up to 30% in growing sika deer. No effect was observed on the growing performance, the feed intake and the hematocrit and hemoglobin level, when SMS was used up to 20%. Higher percentage decreased the level of hematocrit and hemoglobin. On the contrary, immunoglobulin M was increased regardless of the used SMS levels. Furthermore, enzyme gamma glutamyl transferase was increased, an indication that ensiled *Pleurotus ostreatus* SMS can enhance the antioxidant capacity of the animals. Yuan et al. (2022a) also performed a study on male sika deer, using ensiled *Flammulina velutipes* SMS as feed ingredient. No changes were observed regarding growth performance, feed intake, hematological and biochemical parameters, with the exception of immunoglobulin A that was decreased. According to Yuan et al. (2022a), since immunoglobulin A is increased due to external stress stimulation, incorporating *Flammulina velutipes* SMS in the nutrition of deer can contribute to the alleviation of the acute stress that the animals experience when cutting the velvet antler.

The inclusion of mushroom SMS in elk feed has also similar benefits. Park et al. (2012) replaced the standard corn-wheat concentrated feed of elk with ensiled *Pleurotus eryngii* up to 20%. In contrast to the studies on deer (Yuan et al., 2022a; 2022b) hemoglobulin and hematocrit levels were increased. Cholesterol and total triglycerides concentrations were not affected, while high density lipoprotein content and energy intake were increased. Thus, *Flammulina velutipes* SMS was proposed as beneficial not only for the feed cost reduction but also for the health of the animals.

Studies concerning the utilization of mushroom by-products in deer and elks are summarized in Table 1.

Application of Mushroom By-Products in Poultry

Broilers

Mushroom by-products in the diet of broilers have been studied both as a nutritional source and a mean to boost their health status. Although SMS is the main by-product of mushroom cultivation, mushroom waste such as stems and fruiting bodies that do not meet the commercial standards and mycelium are also by-products that could be exploited as livestock feed (Antunes et al., 2023). *Pleurotus ostreatus* SMS (Foluke et al., 2014; Mthana & Mthiyane, 2024), waste (Fard et al., 2014; Hassan et al., 2020) or mycelium powder (Toghyani et al., 2012) have been extensively studied in poultry. In all studies, the SMS or waste that is left after the harvesting of the fruiting powders is gathered, dried, milled and incorporated in the diet of broilers. Foluke et al. (2014) evaluated the replacement of wheat bran by SMS in the diet of broilers. SMS increased both feed intake and weight gain. According to the authors, SMS has a high content of fiber that results in dilution of the other nutrients, and so the birds need to consume higher quantities of feed in order to cover their nutritional needs. In addition, the pasteurization of the SMS before its use as feed ingredient increased its nutritional value and digestibility. Feed conversion ratio was also decreased, especially at the latest phase of birds' growth. Mthana and Mthiyane (2024) replaced soya bean in the diet of broilers by a mixture of *Pleurotus ostreatus* SMS, mucuna seed and marula seed cake. Although replacement by only mucuna seed and maruna seed cake reduced feed intake, growth performance, cholesterol and meat hue angle and increased meat redness, simultaneous addition of SMS with these two feedstuffs ameliorated all the negative effects with the exception of cholesterol that was further increased. It has been observed that although *Cordyceps militaris* SMS decreases cholesterol concentration in the meat of pigs (Boontiam et al., 2020; Qi et al., 2020) and the egg yolk of laying hens (Hwang et al., 2012), *Pleurotus ostreatus* SMS has no effect on the cholesterol level in meat of lamb (Ehtesham and Vakili, 2015) or elk (Park et al., 2012). In accordance with the findings of Foluke et al. (2014), Fard et al. (2014) and Hassan et al. (2020) observed that incorporation of 1% w/w *Pleurotus ostreatus* waste in the diet of broilers increased feed intake and body weight gain. On the other hand, incorporation of 2% w/w had no (Hassan et al., 2020) or negative (Fard et al., 2014) effect on these performance parameters. Furthermore, the use of *Pleurotus ostreatus* waste boosted the immune system of the birds and augmented the antibody titer production after vaccination against Newcastle disease (Fard et al., 2014; Hassan et al., 2020) without affecting the growth of inner organs and increasing cholesterol concentration. Thus, *Pleurotus ostreatus* waste is a valuable ingredient to be examined as an alternative to antibiotics in poultry diets (Fard et al., 2014; Hassan et al., 2020). On the contrary, Toghyani et al. (2012) reported that incorporating *Pleurotus ostreatus* powder in the diet of broilers does not affect their immune response to Newcastle and influenza diseases. Nonetheless, in this study the powder was used as prebiotic and in combination with mannan-oligosaccharides. Hemoglobulin, hematocrit and cholesterol levels were not affected, but serum triglycerides concentrations were increased (Toghyani et al., 2012).

Pleurotus eryngii by-products have also been evaluated as a feedstuff for poultry; SMS after composting (Chuang et al. 2020; 2021), stem left after harvesting, dried and milled to powder (Lee et al., 2012) or co-fermented with soyabean hulls before the incorporation (Lai et al., 2015). The use of composted *Pleurotus eryngii* SMS as feed ingredient for broilers decreases feed conversion ratio either used as feed additive up to 0.5% w/w (Chuang et al., 2020) or as a replacement of soyabean up to 5% w/w in the ration (Chuang et al., 2021). In both cases, not only it exerted antioxidant but also anti-inflammatory activities by decreasing the expression of genes related to inflammation. Thus, SMS ameliorated the negative effects of oxidative stress in the birds. According to Chuang et al. (2020; 2021), SMS displays these properties due to the high content of bioactive phenolic compounds which are contained in the mushroom. Furthermore, *Pleurotus eryngii* SMS stimulated the metabolism of adipose tissue resulting in an increase of serum triglycerides concentration and acted as a prebiotic by enhancing the health of gut barrier and the balance of intestinal microbiota (Chuang et al., 2020). Not only the SMS (Chuang et al., 2020; 2021) but also the stem left after the harvesting of *Pleurotus eryngii* (Lee et al., 2012) can provide antioxidant benefits when added in the diets of broilers by increasing the levels of antioxidant enzymes and decreasing the concentration of MDA. Additionally, SMS has positive results on meat color, capacity of retaining water and drip loss upon storage. Finally, Lai et al. (2015) observed that co-fermentation of *Pleurotus eryngii* SMS with soyabean hulls waste enhanced the phenolic content and the nutritional profile of the mushroom by-product. Addition of the co-fermented mixture at 5% w/w in the diet of broilers improved body weight gain, intestinal morphology, and microbiota.

Another mushroom by-product that has been assessed as a feedstuff for broilers is *Pleurotus sajor-caju* SMS, dried (Adetunji & Adejumo, 2019) or composted (Azevedo et al., 2009). It improves the performance of broilers by increasing daily body weight gain (Adetunji & Adejumo, 2019; Azevedo et al., 2009). Feed intake and feed conversion ratio are not affected when used as composted (Azevedo et al., 2009), but are positively influenced when used dried and milled without any extra composting process (Adetunji & Adejumo, 2019). According to Azevedo et al. (2009), SMS has no effect on carcass abdominal fat, but it increases the villus height.

Stem residues of *Flammulina velutipes* were sun-dried and mixed in the rations of broilers (Mahfuz et al., 2019; 2020). Although no effect on feed intake and body growth was observed, the mushroom by product exerted immunoregulatory activities and enhanced the production of antibody titers following vaccination against Newcastle disease or Avian influenza (Mahfuz et al., 2019). Additionally, cholesterol (Mahfuz et al., 2019; 2020) and total triglycerides (Mahfuz et al., 2020) concentration were decreased. According to Mahfuz et al. (2020), adding *Flammulina velutipes* stem waste in poultry diet decreased MDA level and improved oxidative stability.

Table 2. Application of mushroom by-products in broilers

Species	Form	Level	Results	Reference
Agaricus bisporus	Powder	0.1-0.2%	Improvement of feed efficiency, growth performance (0.2%) and tissue antioxidant-protective activity	Giannenas et al., 2010
Agaricus blazei	SMS	0.2, 0.4, 0.6, 0.8, and 1.0%	Improved weight gain, feed conversion rate and feed intake at 0.2%	Machado et al., 2007
Antrodia Cinnamomea	Powder	0.1, 0.2 or 0.4%	Increased body weight gain, but no effect on feed intake and FCR. Improved antioxidant and anti-inflammatory properties	Lee et al., 2018
Antrodia cinnamomea	Solid-state fermentation with wheat barn	0.5 or 1%	Improved intestinal microflora, antioxidant enzymes activities, meat fatty acid profile and weight gain	Lee et al., 2020
Cordyceps militaris (WM)/ Pleurotus eryngii (SR)/Pleurotus sajorcaju (SR)/Fammulina velutipes (SR)	Combination of waster medium (WM) with stalk residue (SR)	0.5-1.0%	Promotion of growth performance and antioxidant capacity with *Cordyceps militaris* WM showing the best results	Hsieh et al., 2021
Cordyceps militaris	SMS	0.1, 0.25 or 0.5%	No effect on growth performance, reduction of total cholesterol and low-density lipoproteins	Noopan et al., 2019
Flammulina velutipes	Stem	1 or 2%	Increase of serum immunoglobulin G, antibody titers against Newcastle disease (ND) and infectious bursal disease (IBD) (2%) and decrease of total cholesterol levels. No effect on body weight, feed intake and FCR.	Mahfuz et al., 2019
Flammulina velutipes	Stem	1 or 2%	Decrease of total cholesterol, triglycerides and MDA content in serum, thigh, breast and liver	Mahfuz et al., 2020
Pleurotus eryngii	Stem	0.1, 0.5, 1 or 2%	Positive effects on meat color and water holding capacity, decreased fat and MDA content, and increased activities of antioxidative enzymes	Lee et al., 2012

Species	Form	Level	Results	Reference
Pleurotus eryngii	Co-fermented stalk residues and soybean hulls by Aureobasidium pullulans	0.5 or 1%	Increase of body weight gain, improvement of intestinal morphology and microbiota	Lai et al., 2015
Pleurotus eryngii	Pennisetum purpureum Schum waste mushroom compost	0.5, 1 or 2%	Decrease of FCR (0.5%), glucose, triglyceride levels (0.5 and 2%), IFN-γ and IL-1ß, increase of adiponectin, oxytocin, Nrf-2 and SOD-1	Chuang et al., 2020
Pleurotus ostreatus	Powder	0.1 or 0.2%	Increase of body weight (0.2%) and decrease of serum triglyceride concentration. No effect on carcass yield, internal organs relative weight, antibody titer production against Newcaste and influenza, and hematological and biochemical parameters	Toghyani et al. 2012
Pleurotus ostreatus	Waste (discarded fruiting bodies and stems)	1 or 2%	Increase of feed intake and weight gain with 1% but not 2%, decrease of H/L ratio and MDA levels	Hassan et al. 2020
Pleurotus ostreatus	Waste (discarded fruiting bodies and stems)	1 or 2%	No effect on feed intake, body weight and weight gain with 1% but 2% has negative effect on BW and FCR, increase of jejunal villus height and crypt depth and decrease of H/L ratio	Fard et al. 2014
Pleurotus ostreatus	SMS	Replacement of wheat bran at 25, 50, 75 and 100%	Increase of feed intake and weight gain, no effect on carcass traits	Foluke et al. 2014
Pleurotus ostreatus	SMS combined with marula seed cake and mucuna seed meal to replace soya bean meal	1.25, 2.5 or 5%	Improvement of feed intake, growth performance, carcass weight and meat quality (1.25%) and decrease in serum alkaline phosphatase and cholesterol	Mthana & Mthiyane 2024
Pleurotus sajor-caju	SMS (standard, mutant and wild)	6%	No effect on body weight gain and FCR. Decrease of feed intake	Adetunji & Adejumo 2019
Pleurotus sajor-caju	SMS	0.5, 1, 1.5 or 2%	No effect on FCR, feed intake, carcass yield and abdominal fat. Increase of body weight gain	Azevedo et al. 2009

Table 3. Application of mushroom by-products in laying hens, geese, pigs and rabbits

Species	Mushroom strain	Form	Level	Results	Reference
Laying hens	*Cordyceps militaris*	SMS	0.5, 1 or 2%	Decrease of egg cholesterol and feed conversion ratio (2%), increase of egg mass and no effect on yolk color and eggshell thickness	Wang et al. 2015
Laying hens	*Flammulina velutipes*	Mycelium fermented with *Bacillus subtilis* A8-8 and *Klebsiella sp.* Sc	1, 2, 3, 4 or 5%	No effect on egg production, feed intake and feed conversion ratio, increase of egg weight (1, 3%) albumen height, Haugh unit, egg-shell weight and shell thickness (4%), and decrease of Escherichia coli and Salmonella spp. in caecum and ammonia production (3-5%)	Lee et al. 2014
Laying hens	*Flammulina velutipes*	Stem	2, 4 or 6%	No effect on average daily feed intake, average daily weight gain, feed conversion ratio and internal organ weights, apart from proventriculus and bursa that were increased. Increase of antibody titres against Newcastle disease, infectious bronchitis and avian influenza virus and serum IgA, IgG, IgM.	Mahfuz et al. 2018
Laying hens	*Flammulina velutipes*	Stem	2,4 or 6%	Increase of serum (6%) and yolk (4%) total antioxidant capacity and serum glutathione peroxidase and decrease of serum and yolk MDA.	Chen et al. 2020
Laying hens	*Hypsizygus marmoreus*	Fermented SMS with *Bacillus subtilis*	5, 10 or 15%	No effect on egg production, egg weight, egg mass, feed conversion ratio, eggshell breaking strength, thickness and haugh unit, while yolk color was more intense	Kim et al. 2014
Laying hens	*Lentinula edodes*	Powder	0.25 or 0.5%	Increase of egg production, Haugh unit, total n-6 and polyunsaturated fatty acids, decrease of eggshell thickness and cholesterol (0.5%)	Hwang et al. 2012
Geese	*Pleurotus ostreatus*	SMS	5, 10 or 15%	No effect on FCR, feed intake and hematological parameters, decrease of MDA levels (15%) and positive effect on meat color, flavor and acceptability score	Chang et al. 2016

Species	Mushroom strain	Form	Level	Results	Reference
Growing pigs	*Cordyceps militaris*	SMS	0.2%	Increase of final body weight, average daily weight gain, IgA, IgG, total antioxidant capacity and glutathione peroxidase activity and decrease of leukocyte percentage, cholesterol and MDA. No effect on average daily feed intake, gain-to-feed ratio, glucose, aspartate aminotransferase, triglyceride, high-density lipoprotein, and low-density lipoprotein	Boontiam et al. 2020
Weaned piglets	*Cordyceps militaris*	Fermented SMS	3%	No effect on final weight, average daily gain, and feed conversion ratio. Increase of IgA and IgG, improvement of intestinal mucosal barrier and composition of the microbiota and decrease of serum total protein, albumin, total cholesterol, and total triglyceride	Qi et al. 2020
Fattening pigs	*Pleurotus eryngii*	Fermented SMS with soybean meal and corn	20, 50 or 80%	Improvement of protein content, water holding capacity and tenderness and decrease of volatile basic nitrogen	Chu et al. 2011
Finishing pigs	*Pleurotus ostreatus*	Fermented SMS with rice bran and barley bran	3, 5 or 7%	Improvement of average daily feed intake, growth performance, carcass traits, meat quality and fatty acid profile at 3%	Song et al. 2007
Fattening rabbits	*Not mentioed*	SMS	20, 40 or 60% of berseem hay	No effect on average daily gain and total feed intake. Improvement of final body weight (60%), nutrient digestibility and total digestible nutrient value (40-60%) and decrease of pH and NH3-N concentrations	Morshedy et al. (2023)
Fattening rabbits	*Pleurotus ostreatus*	SMS	20000, 40000 or 60000 IU/kg of laccase enzymes	Decrease of total weight gain, feed intake, daily weight gain, hot carcass weight, and dressing percentage as the amount of enzyme increased.	Saavedra-Castillo et al. (2023)
Fattening rabbits	*Pleurotus ostreatus*	SMS	Up to 20% of mulberry leaves	No effect was observed on the final weight, daily weight gain, feed intake and feed conversion ratio.	Martínez Ramírez et al. 2018

Cordyceps militaris stalk waste is another mushroom by-product that can be exploited in the diet of broilers (Noopan et al., 2019) with similar results as these of *Flammulina velutipes* (Mahfuz et al., 2019; 2020). Although it does not affect the performance of poultry, it has a positive impact on lipid metabolism by increasing total cholesterol and low-density lipoproteins. Lee et al. (2014) observed that dietary addition of 1% w/w of *Cordyceps militaris* waste medium led to the highest growth performance and antioxidant response compared to adding 0.5% w/w *Cordyceps militaris* waste medium mixed with 0.5% w/w *Pleurotus eryngii* or *Pleurotus sajor-caju* or *Flammulina velutipes* stem residues. Thus, the authors recommended *Cordyceps militaris* as the most promising mushroom species that can be used as feed additive in broilers and should be further examined.

Another strain that has been tested as poultry feed additive is *Antrodia cinnamomea* (Lee et al., 2019; 2020). Lee et al. (2018) added *Antrodia cinnamomea* mycelium powder in the diet of broilers and although feed intake was increased, feed conversion ratio remained unaffected. At the same time, the powder exerted immunoregulatory and anti-inflammatory activities. In another study, Lee et al. (2020) cultivated *Antrodia cinnamomea* using solid state culture on wheat barn, and the fermented product was added as powder in the diet of broilers. It enhanced the growth and the production of antioxidant enzymes and altered the tissue lipid profile by increasing the level of monounsaturated fatty acids and reducing the level of saturated fatty acids. Based on the above findings, Lee et al. (2020) state that mushroom by-products can be applied as a mean to alter tissue lipid profile.

The literature regarding the utilization of mushroom by-products in broilers is summarized in Table 2.

Laying Hens

As in broilers, there are various by-products that have been studied as feed ingredients in the diets of laying hens; mushroom stems dried and milled (Chen et al. 2020; Mahfuz et al., 2018), mushroom mycelium powder (Lee et al., 2014; Hwang et al., 2012), and SMS as left after mushroom harvesting (Wang et al., 2015) or followed by an extra fermentation that enhance nutritional value (Kim et al., 2014). The main mushroom strain that has been researched is *Flammulina velutipes* as stem (Chen et al., 2020; Mahfuz et al., 2018) and mycelium powder (Lee et al., 2019). Addition up to 6% w/w of *Flammulina velutipes* stem by-product in the diet of ISA Brown layers exerts antioxidant beneficial effects on both blood serum and egg yolk, even higher than adding an antibiotic (flavomycin) in the diet, due to the phenolic compounds included in the mushroom (Chen et al., 2020). Additionally, Mahfuz et al. (2018) observed that incorporating 6% w/w *Flammulina velutipes* stem in the rations of laying hens of the same hybrid, increased the calcium deposited on the eggshell and the general feed calcium utilization, although it did not affect the feed intake, feed conversion ratio, and egg production. Moreover, it resulted in higher antibody titers against Newcastle disease and infectious bronchitis, as observed also in broilers (Fard et al., 2014; Hassan et al., 2020), and strengthened the immune system of the birds by increasing immunoglobulin A and G compared to flavomycin. In accordance with Mahfuz et al. (2018), Lee et al. (2014) state that adding *Flammulina velutipes* mycelium powder in the diet of laying hens does not increase feed intake and egg productivity, but it does improve egg traits. Albumen height and eggshell thickness are increased but yolk color is not affected. The increase of eggshell thickness may be attributed to the high calcium deposition, as observed by Mahfuz et al. (2018). The use of *Flammulina velutipes* stem has also antimicrobial effects through the suppression of pathogenic bacteria (*Escherichia coli* and *Salmonella spp.*) in the gastrointestinal system, by acting as prebiotic (Mahfuz et al., 2018).

Although addition of *Flammulina velutipes* powder (Lee et al., 2014) or stem (Mahfuz et al., 2018) in the diet of laying hens does not affect egg characteristics with the exception of increasing eggshell thickness, adding *Cordyceps militaris* SMS increases cell mass and egg white (Wang et al., 2015), while *Hypsizygus marmoreus* fermented SMS improved yolk color according to Kim et al. (2014). *Hypsizygus marmoreus* fermented SMS also decreases feed conversion ratio (Kim et al., 2014). Nonetheless, neither *Cordyceps militaris* (Wang et al., 2015) nor *Hypsizygus marmoreus* (Kim et al., 2014) SMS affect the eggshell thickness in contrast to *Flammulina velutipes* powder (Lee et al., 2014) or stem (Mahfuz et al., 2018) that increase it. Furthermore, *Cordyceps militaris* SMS reduces the concentration of cholesterol in the eggs (Wang et al., 2015).

It has also been observed that using *Lentinula edodes* powder in the diet of laying hens decreases egg yolk cholesterol level (Hwang et al., 2012), although egg productivity and eggshell thickness were reduced. Hwang et al. (2012) observed also that the fatty acid composition of the mushroom by-product affected the fatty acid composition of the egg. Since *Lentinula edodes* is rich in linoleic acid, the egg yolk has also increased levels of linoleic acid, n-6, and polyunsaturated fatty acids. Thus, Hwang et al. (2012) highlight that mushrooms used as feed additives could be a mean to affect not only cholesterol level but also fatty acid profile in the yolk, as proposed by Lee et al. (2020) for the tissue lipid profile.

Studies concerning the utilization of mushroom by-products in laying hens are summarized in Table 3.

Geese

Regarding other poultry species, mushroom by-products have also been evaluated in the diets of geese. Incorporating composted *Pleurotus ostreatus* SMS in the diet of White Roman geese resulted in a reduction of the weight gain, although not affecting feed conversion ratio and feed intake (Chang et al., 2016). Still, it showed antioxidant activities and positive effect on the color, flavor, and the acceptability score of the meat (Table 3).

Application of Mushroom By-Products in Pigs

SMS has been evaluated as an ingredient of pig diets not only because of the nutrients that it contains, but also due to its bioactive compounds that induce immunoregulatory and antimicrobial effects. Three strains have been examined in pigs, *Cordyceps militaris* (Boontiam et al., 2020; Qi et al., 2020), *Pleurotus eryngii* (Chu et al., 2011) and *Pleurotus ostreatus* (Song et al., 2007). In all studies, SMS was used after fermentation, with the exception of the study by Boontiam et al. (2020). Boontiam et al. (2020) harvested SMS of *Cordyceps militaris*, dried and incorporated it into the ration of growing pigs. The percentage of SMS in the ration was 0.2% w/w. Although the average daily feed intake was not affected, the final body weight and average daily weight gain of the animals were increased. Additionally, immunoglobulin A and G were increased, thus the authors proposed the use of *Cordyceps militaris* in growing pigs' diets for inducing immunostimulatory effects. It was also observed that cholesterol concentration was decreased, and so SMS was proposed as a mean of controlling hyperlipidemia. SMS induced also malondialdehyde (MDA) levels reduction, a substance that is produced after lipid oxidation, and as a result it was concluded that SMS also provides antioxidant benefits. Qi et al. (2020) incorporated fermented *Cordyceps militaris* SMS in the diet of weaned piglets. In this experiment the authors were particularly interested in the immunological activity of the mushroom by-products because weaning is stressful for the animals since at this age piglets have still a sensitive immune system. The percentage of fermented SMS in the ration was 3% w/w. No change was observed regarding weight gain and growth performance, in contrast to the study of Boontiam et al. (2020). In accordance with Boontiam et al. (2020), immunoglobulin A and G were increased, and total cholesterol and triglycerides were decreased. Qi et al. (2020) also observed that the fermented SMS acted as a probiotic, altering the microbiome in the colon and the cecum, and benefiting the mucosal barrier.

SMS generated by *Pleurotus eryngii* (Chu et al., 2011) and *Pleurotus ostreatus* (Song et al., 2007) has been incorporated in higher percentage in pigs' ration compared to that of *Cordyceps militaris* (Boontiam et al., 2020; Qi et al., 2020) after mixture and fermentation with other feedstuffs. Song et al. (2007) mixed *Pleurotus ostreatus* with rice and barley bran and then fermented the mixture, which was then evaluated in the diet of finishing pigs at the levels of 3, 5 and 7% w/w. 3% w/w increased the daily feed intake, the carcass weight, and the crude protein and decreased the feed convention ratio. On the other hand, 5 and 7% w/w had the opposite effect. The meat quality was also affected. In all cases, cooking loss and water holding capacity was increased, pH was decreased but color was not affected. Moreover, Song et al. (2007) observed that the SMS affected the fatty acid profile of pork meat, with myristic and palmitoleic acid being higher in the 7% w/w group and palmitic and stearic acid being lower in the meat of pigs that were raised with no SMS in their ration. Based on the above observations, Song et al. (2007) conclude that 3% fermented SMS of king oyster mushroom can be beneficial for growth performance and carcass characteristics and that can be studied as a mean to beneficially manipulate pork meat fatty acids composition. Chu et al. (2011) evaluated the addition of fermented *Pleurotus eryngii* mixed with brewer's grain and rice barn in the diet of fattening pigs from 20% to 100% of their total ration. This is the highest percentage of SMS that has been studied as an ingredient for pig diets. The authors used *Lactobacillus planetarium*, *Enterococcus faecium* and *Saccharomyces cerevisiae* as probiotics for the fermentation in order to increase crude protein levels and lactic acid content and decrease pH values. The study of Qi et al. (2020) used *Bacillus subtilis* for the fermentation but the study of Boontiam et al. (2020) does not refer to the microorganism. Different microorganisms that induce the fermentation lead to different metabolites and by-products, so they can affect the final nutritional value and bioactive content of the fermented SMS. Consequently, it is difficult to compare the results of the studies. This will be also analyzed in the discussion part of this review. Chu et al. (2011) prepared a fermented mixture of the SMS with grains, of a high crude protein content, near to 20%. In contrast to Song et al. (2007), they observed reduction of feed intake and growth performance, but improved carcass traits. Chu et al. (2011) also mention that the composted SMS decreased the feed cost, so its application in pig diets can provide economical, in addition to environmental and health, benefits.

In addition to its use as feed ingredient for pig diets, SMS has been valorized also as environmental enrichment. Beattie et al. (2001) have studied the use of mushroom compost as substrate that can promote rooting behavior and enhance the animals' welfare, without referring to mushroom species. Mushroom compost was placed in a rank over pigs' head. Rooting behavior was increased, and tail biting was decreased, so the authors proposed the use of mushroom by-products as a mean to promote rooting and decrease aggressive behavior.

The literature regarding the utilization of mushroom by-products in pigs is summarized in Table 3.

Application of Mushroom By-Products in Rabbits

Morshedy et al. (2023) replaced berseem hay with SMS in the diet of fattening rabbits up to 60%. The chemical analysis indicated that SMS had higher ash content and cell wall constituents (polysaccharides) but the content in the other nutrients was lower than in berseem hay. Nonetheless, using 60% SMS instead of berseem hay in the diet of the rabbits resulted in improved final body weight and feed conversion ratio, although the total feed intake was not altered. Moreover, 40% and 60% SMS improved feed digestibility. Martínez Ramírez et al. (2018) incorporated milled *Pleurotus ostreatus* SMS in the diet of rabbits. In this study, the rabbits were fed a diet consisting of soybean meal, maize, sugar cane and mulberry leaves and SMS substituted the mulberry leaves up to 20%. No effect was observed on the final weight, daily weight gain, feed intake and feed conversion ratio. The effects on meat characteristics were not studied. Morshedy et al. (2023) and Saavedra-Castillo et al. (2023) both suggested SMS as a cost-effective feed additive for rabbits that should be further evaluated. Saavedra-Castillo et al.

(2023) studied enzyme extract of *Pleurotus ostreatus* SMS. The extract contained cellulases, xylanases and laccases, which degrade lignocellulosic substrates and at the same time increase crude protein levels. It was assumed that adding this extract to the diet of rabbits would increase digestibility. On the contrary, total weight gain and feed intake were decreased as the concentration of enzymes was increased in the ration. Carcass traits and meat characteristics were not affected when the laccases enzymes were added at 40000 IU/kg, although meat cooking loss and hardness were increased. The authors concluded that although further studies are required, *Pleurotus ostreatus* SMS can be a feed additive beneficial for both minimizing feed cost and improving feed digestibility (Saavedra-Castillo et al., 2023).

Studies concerning the utilization of mushroom by-products in rabbits are summarized in Table 3.

Discussion

The research on exploiting mushroom by-products in livestock diets is growing. As indicated by the results of the present review study, the research until today has focused mainly on broilers, but other monogastric and ruminant species have also been studied. Our results indicate that the main mushroom species that have been utilized are *Pleurotus ostreatus*, followed by *Flammulina velutipes* and *Pleurotus eryngii*. The main mushroom by-product evaluated in livestock diets is the spent mushroom substrate (SMS), but evaluation of stem waste is also gaining increasing attention in poultry. No results have been retrieved about the use of stem by-products as ingredient in the diet of other livestock species. The findings on using stem as feedstuff for poultry demonstrate that it has beneficial effects on both animals' performance and health status. It promotes growth performance in broilers (Lai et al., 2015; Mahfuz et al., 2018), calcium deposition of eggshell in laying hens (Mahfuz et al., 2018), has antioxidant effects (Chen et al., 2020; Hsieh et al. 2021; Mahfuz et al. 2020) and immunostimulatory properties (Mahfuz et al., 2018; 2019; 2020) in both broilers and laying hens. These are promising results, hence, further research on the use of mushroom stem by-products in livestock diets is warranted.

SMS is rich in cellulose, hemicellulose, and lignin. The fermentation process includes enzymatic degradation of these substances that improves their digestibility and increases the substrate's content in nutrients (Chuang et al., 2020). SMS contains the proteins of the mushroom along with proteins produced by the fermentation process, and is abundant in triterpenes, polysaccharides, glycoproteins, polyphenols, and other high-value substances contained in the mushrooms (Antunes et al., 2020; Baptista et al., 2023; Chuang et al., 2020). As indicated, it has been evaluated as raw in the diet of animals or after biological and other processes that can augment its nutritional content and digestibility even further. Foluke et al. (2014) applied pasteurization and reported that the thermal treatment increased digestibility. Composting (Chuang et al., 2020; Ehtesham and Vakili, 2015) and ensilaging (Seok et al., 2016; Zaboli et al. 2023) have also been used in various studies with similar results. Additionally, SMS can be inoculated with various bacteria to enhance different properties, depending on the selected microorganism (Kim et al., 2012; Lai et al., 2015; Lee et al., 2014). Higher nutritional value can be also achieved when co-treated with other materials (e.g. molasses, Zaboli et al. 2023). The aforementioned possibilities demonstrate that SMS can be a valuable ingredient in livestock diets, since it can undergo various processes that would beneficially affect its properties. Moreover, inoculation with bacteria can decrease its moisture content and pH and so prolong its storage time. Baek et al. (2017) managed to expand the storage time of *Pleurotus ostreatus* SMS after inoculation with *Lactobacillus brevis* for five days and Kim et al. (2011) after incoculation with a mixture of *Lactobacillus plantarum Lp1', Pediococcus acidilactici Pa193, L. plantarum Lp2M* for 17 days.

As analyzed above, edible mushrooms have antioxidant properties (Ritota and Manzi, 2019). Additionally, adding SMS to an animal's diet provides benefits for its health as well as to its nutrition. It can serve as an anti-inflammatory factor by boosting the immune system and improving microbiological resistance, as an antioxidant by raising the concentration of antioxidant enzymes and assisting in the reduction of lipid peroxidation, and as probiotic by supporting the balance of intestinal microflora. As a result, it might help in the reduction of the antibiotics' use. SMS contains compounds with immunoregulatory and antimicrobial properties (Boontiam et al., 2020; Qi et al., 2020; Chu et al., 2011; Shong et al., 2007). Studies have found that incorporation of *Pleurotus ostreatus* SMS in broiler diets, enhance antibody titer production, and have immunomodulatory properties (Fard et al., 2014; Hassan et al., 2020). Same results have also been observed in laying hens for *Flammulina velutipes* stem (Lee et al., 2020). In all studies, the mushroom by-products have performed better than the used antibiotics. Moreover, Liu et al. (2015b) conclude that *Ganoderma balabacense* extract can decrease somatic cell count in goat milk, although in this study the performance of the mushroom by-product is not compared to any antibiotic. Multiple resistances to traditional antimicrobials in animal feed formulations raise concerns, encouraging research on natural alternatives, hence, mushroom by-products could be utilized in animal husbandry as an alternative.

According to the literature, various studies demonstrate the antioxidant properties of mushroom by-products. Boontiam et al. (2020) found that the incorporation of *Cordyceps militaris* SMS in the ration of growing pigs has an antioxidant effect through the reduction of MDA levels. Composted *Pleurotus eryngii* SMS in the diet of broilers acts as antioxidant and anti-inflammatory agent by decreasing inflammation-related gene expression (Chuang et al., 2020). Both *Pleurotus eryngii* SMS and stem can improve broiler health status by increasing the concentration of antioxidant enzymes and reducing MDA and lipid peroxidation (Chuang et al., 2020; 2021; Lee et al., 2012). *Flammulina velutipes* stem by-product results in antioxidant effects on blood serum and egg yolk in laying hens, that are more intense than that induced by antibiotics (Chen et al., 2020). Addition of composted *Pleurotus ostreatus* SMS in white Roman geese diets enhances antioxidant properties, while *Ganoderma lucidum* and *Ganoderma balabacense* water extracts increase immunoglobulin A (Liu et al., 2015a) and serum antioxidant capacity (Liu et al., 2015b). Ensiled *Pleurotus ostreatus* SMS can replace standard concentrated supplements up to 30% w/w in sike deer diets and simultaneously enhance the animals' defense to oxidative stress (Yuan et al. 2022a). Antioxidant activities of mushroom by-products is another benefit of their use in livestock diets and another reason to continue the respective research.

Apart from their antioxidant and immunostimulatory effects, mushroom by-products could be also used as a mean to alter the lipid profile in both meat and egg, since the fatty acid composition of the mushroom by-product can affect the fatty acid composition of the animal products. Lee et al. (2020) demonstrated that fermented *Antrodia cinnamomea* as an ingredient in the diet of broilers increased the level of monounsaturated fatty acids and reduced the level of saturated fatty acids. Liu et al. (2015b) observed that *Ganoderma balabacense* extract increased total triglycerides and protein level in the milk of goats. Shong et al. (2007) found that *Pleurotus ostreatus* SMS altered the fatty acid profile in pork meat and Hwang et al. (2012) observed that incorporation of *Cordyceps militaris* SMS into the diet of layers resulted in decreased egg yolk cholesterol level. Additionally, studies report that incorporation of mushroom by-products in the diet of pigs (Boontiam et al., 2020; Qi et al., 2020) and broilers (Mahfuz et al., 2019; Mahfuz et al., 2020) decreases cholesterol and total triglycerides concentration. Therefore, mushroom by-products can be used as a mean to both enhance the health status of animals and alter the lipid profile of the final product to meet consumer demands.

Studies have also shown that mushroom by-products improve carcass traits and meat quality characteristics. *Pleurotus eryngii* stem enhanced meat color and water-holding capacity in broilers (Lee et al., 2012). *Pleurotus eryngii* SMS also increased the rib eye area and improved carcass traits in calves (Kim et al., 2012). Furthermore, it improved meat water-holding capacity without affecting color and flavor in geese meat (Chang et al., 2016). Finally, *Pleurotus eryngii* SMS increased rib eye area, backfat thickness, cold and hot carcass weight in lambs (Aldoori et al., 2015). The effect of mushroom by-products on meat traits could also be a proposal for future research.

Furthermore, various literature findings support the enhancement of growth performance when using mushroom by-products in cattle (Kime et al., 2011), sheep (Huang et al., 2023), pigs (Qi et al., 2020), rabbits (Morshedy et al., 2023), broilers (Hsieh et al., 2021) and geese (Chang et al., 2016). Moreover, it increases milk yield when used in the diet of dairy cattle (Liu et al., 2015b) and goats (Fan et al., 2022). Regarding egg production, although it does not increase it, it improves eggshell thickness, albumen height (Lee et al., 2014) and egg mass (Wang et al., 2015) and decreases feed conversion ratio (Wang et al., 2015). Thus, mushroom by-products can be a cost-effective feedstuff that enhance simultaneously animal product quality. Studies that support this cost-effective result by cost benefit analysis, while simultaneously demonstrating the positive effect on productivity and product quality could persuade the farmers and stakeholders to incorporate mushroom by-products in the diet of livestock animals. Among all the literature findings, only Mandale et al. (2023) calculated the cost reduction, while proving the positive effect of *Agaricus bisporus* SMS on the growth performance and immune response of goats. The economic analysis of using mushroom by-products as ingredient for livestock is out of the scope of our review and no separate research has been conducted to examine if there are studies that focus exclusively on it. Nonetheless, we propose that studies that calculate the cost reduction while demonstrating the benefits on productivity and health status could be useful as a future direction.

A further direction for future studies should also be the evaluation of SMS as substrate for environmental enrichment. Beattie et al. (2001) studied the use of mushroom compost as a substrate to promote rooting behavior in pigs and enhance animal welfare. Exploration was promoted, indicator of positive welfare, while tail biting was decreased. In addition to its use in livestock nutrition, SMS could be used as substrate to promote natural behaviors such as comfort and exploration in livestock species.

Conclusion and Future Perspectives

Global mushroom production has been rapidly increased in recent decades due to the demand for high quality food with minimal environmental impact. As a result, there is a rise in the produced SMS quantities that are a potential environmental pollutant if they are not appropriately handled. The huge amount of SMS can be handled in various ways, but they should be utilized in a circular way to improve environmental friendliness and economic viability for a sustainable mushroom production sector.

Valorizing SMS is crucial for developing a sustainable mushroom industry in a circular-economy model. Mushroom cultivation produces nutritious final products, but by-products like mycelium, stem, and SMS remain. SMS has been evaluated as a feed ingredient for livestock diets. Incorporating SMS in animals' nutrition can improve their health status by acting as a probiotic, antioxidant, and anti-inflammatory factor, promoting intestinal microflora balance, thus reducing antibiotic use. Furthermore, SMS could be used in the diets of ruminants and monogastric animals, as it has been found to improve animal health and productivity, since it possesses antibacterial, cytostatic, antioxidant, and immunomodulatory properties. Finally, SMS could be used in a circular economy model, but quantitative models are needed to predict its environmental impact.

Disclaimer

None

References

Abdel-Aziz NA, Salem AZ, El-Adawy MM, Camacho LM, Kholif AE, Elghandour MM, Borhami BE. Biological treatments as a mean to improve feed utilization in agriculture animals—An overview. *Journal of Integrative Agriculture* (2015) 14(3): 534-543.

Adetunji CO, Adejumo IO. Potency of agricultural wastes in mushroom (*Pleurotus sajor-caju*) biotechnology for feeding broiler chicks (Arbor acre). *International Journal of Recycling of Organic Waste in Agriculture* (2019) 8: 37-45.

Aldoori ZT, Al-Obaidi ASA, Abdulkareem AH, Abdullah MKH. (2015). Effect of dietary replacement of barley with mushroom cultivation on carcass characteristics of Awassi lambs. *Journal of Animal Health and Production* (2015) 3(4): 94-98.

Antunes F, Marçal S, Taofiq O, Morais A, Freitas AC, Ferreira I, Pintado M. Valorization of mushroom by-products as a source of value-added compounds and potential applications. *Molecules* (2020) 25(11): 2672.

Azevedo RS, da Silva Avila CL, Dias ES, Bertechini AG, Schwan RF. Utilization of the spent substrate of *Pleurotus sajor caju* mushroom in broiler chicks ration and the effect on broiler chicken performance/Utilizacao do composto exaurido de *Pleurotus sajor caju* em racoes de frangos de corte e seus efeitos no desempenho dessas aves. *Acta Scientiarum Animal Sciences* (2009) 31(2): 139-145.

Baba E, Uluköy G, Öntaş, C. Effects of feed supplemented with *Lentinula edodes* mushroom extract on the immune response of rainbow trout, Oncorhynchus mykiss, and disease resistance against Lactococcus garvieae. *Aquaculture* (2015) 448: 476-482.

Baek YC, Kim MS, Reddy KE, Oh YK, Jung YH, Yeo JM, Choi H. Rumen fermentation and digestibility of spent mushroom (*Pleurotus ostreatus*) substrate inoculated with Lactobacillus brevis for Hanwoo steers. *Revista Colombiana de Ciencias Pecuarias* (2017) 30(4): 267-277.

Beattie VE, Sneddon IA, Walker N, Weatherup RN. Environmental enrichment of intensive pig housing using spent mushroom compost. *Animal Science* (2001) 72(1): 35-42.

Boontiam W, Wachirapakorn C, Wattanachai, S. Growth performance and hematological changes in growing pigs treated with *Cordyceps militaris* spent mushroom substrate. *Veterinary World* (2020) 13(4): 768.

Chang SC, Lin MJ, Chao YP, Chiang CJ, Jea YS, Lee TT. (2016). Effects of spent mushroom compost meal on growth performance and meat characteristics of grower geese. *Revista Brasileira de Zootecnia* (2016) 45: 281-287.

Chen M, Mahfuz S, Cui Y, Jia L, Liu Z, Song H. The antioxidant status of serum and egg yolk in layer fed with mushroom stembase (*Flammulina velutipes*). *Pakistan Journal of Zoology* (2020) 52(1): 389.

Chu GM, Kang SN, Kim HY, Ha JH, Kim JH, Jung MS, Ha JW, Lee SD, Jin SK, Kim IS, Shin DK Song YM. Effect of substitution of fermented king oyster mushroom by-products diet on pork quality during storage. *Food Science of Animal Resources* (2012) 32(2): 133-141.

Chuang WY, Liu CL, Tsai CF, Lin WC, Chang SC, Shih HD, Shy YM, Lee TT. Evaluation of waste mushroom compost as a feed supplement and its effects on the fat metabolism and antioxidant capacity of broilers. *Animals* (2020) 10(3): 445.

Chuang WY, Lin LJ, Shih HD, Shy YM, Chang SC, Lee TT. Intestinal microbiota, anti-inflammatory, and anti-oxidative status of broiler chickens fed diets containing mushroom waste compost by-products. *Animals* (2021) 11(9): 2550.

Cohen R, Persky L, Hadar Y. Biotechnological applications and potential of wood-degrading mushrooms of the genus *Pleurotus*. *Applied Microbiology and Biotechnology* (2002) 58: 582-594.

Ehtesham S, Vakili AR. The effect of spent mushroom substrate on blood metabolites and weight Gain in Kurdish male lambs. *Entomology and Applied Science Letters* (2015) 2(1): 29-33.

Facchini JM, Alves EP, Aguilera C, Gern RMM, Silveira MLL, Wisbeck E, Furlan SA. Antitumor activity of *Pleurotus ostreatus* polysaccharide fractions on Ehrlich tumor and Sarcoma 180. *International Journal of Biological Macromolecules* (2014) 68: 72-77.

Fan GJ, Chen MH, Lee CF, Yu B, Lee TT. Effects of rice straw fermented with spent *Pleurotus sajor-caju* mushroom substrates on milking performance in Alpine dairy goats. *Animal Bioscience* (2022) 35(7): 999.

FAO. The State of Food and Agriculture. In: *Moving Forward on Food Loss and Waste Reduction*, Rome, Italy, 2019.

FAOSTAT. Available online: http://www.fao.org/faostat/en/#data/QP (accessed on 25 February 2024)

Fard SH, Toghyani M, Tabeidian SA. Effect of oyster mushroom wastes on performance, immune responses and intestinal morphology of broiler chickens. *International Journal of Recycling of Organic Waste in Agriculture* (2014) 3: 141-146.

Foluke A, Olutayo A, Olufemi A. Assessing spent mushroom substrate as a replacement to wheat bran in the diet of broilers. *American International Journal of Contemporary Research* (2014) 4(4): 178-83.

Gerrits JPG. Composition, use and legislation of spent mushroom substrate in the Netherlands. *Compost Science & Utilization* (1994) 2(3): 24-30.

Giannenas I, Pappas IS, Mavridis S, Kontopidis G, Skoufos J, Kyriazakis I. Performance and antioxidant status of broiler chickens supplemented with dried mushrooms (*Agaricus bisporus*) in their diet. *Poultry Science* (2010) 89(2): 303-311.

Hassan RA, Shafi ME, Attia KM, Assar MH. Influence of oyster mushroom waste on growth performance, immunity and intestinal morphology compared with antibiotics in broiler chickens. *Frontiers in Veterinary Science* (2020) 7: 333.

Hatvani N, Kredics L, Antal Z, Mécs I. Changes in activity of extracellular enzymes in dual cultures of *Lentinula edodes* and mycoparasitic *Trichoderma* strains. *Journal of Applied Microbiology* (2002) 92(3): 415-423.

Hsieh YC, Lin WC, Chuang WY, Chen MH, Chang SC, Lee TT. Effects of mushroom waster medium and stalk residues on the growth performance and oxidative status in broilers. *Animal Bioscience* (2021) 34(2): 265-275.

Huang LQ, Yan JC, Yuan ZZ, Li SQ, Tang J. Effects of whole plant rice and spent mushroom substrate of *Pleurotus ostreatus* co-fermented feed on growth performance, nutrient apparent digestibility and blood indices of Liuyang black goats. *Chinese Journal of Animal Nutrition* (2022) 34: 5156-5167.

Huang X, Zhou L, You X, Han H, Chen X, Huang X. Production performance and rumen bacterial community structure of Hu sheep fed fermented spent mushroom substrate from *Pleurotus eryngii*. *Scientific Reports* (2023) 13(1): 8696.

Hwang JA, Hossain ME, Yun DH, Moon ST, Kim GM, Yang CJ. Effect of shiitake [*Lentinula edodes (Berk.) Pegler*] mushroom on laying performance, egg quality, fatty acid composition and cholesterol concentration of eggs in layer chickens. *Journal of Medicinal Plants Research* (2012) 6(1): 146-153.

Kamarudzaman AN, Chay TC, Amir A, Talib S.A. Biosorption of Mn (II) ions from aqueous solution by *Pleurotus* spent mushroom compost in a fixed-bed column. *Procedia-Social and Behavioral Sciences* (2015) 195: 2709-2716.

Kim MK, Lee HG, Park JA, Kang SK, Choi YJ. Recycling of fermented sawdust-based oyster mushroom spent substrate as a feed supplement for postweaning calves. *Asian-Australasian Journal of Animal Sciences* (2011) 24(4): 493-499.

Kim YI, Lee YH, Kim KH, Oh YK, Moon YH, Kwak WS. Effects of supplementing microbially-fermented spent mushroom substrates on growth performance and carcass characteristics of Hanwoo steers (a field study). *Asian-Australasian Journal of Animal Sciences* (2012) 25(11): 1575-1581.

Kim SC, Moon YH, Kim HS, Kim HC, Kim JO, Cheong JC, Cho SJ. Effect of dietary fermented spent mushroom (*Hypsizygus marmoreus*) substrates on laying hens. *Journal of Mushroom* (2014) 12(4): 350-356.

Kitzberger CSG, Smânia Jr A, Pedrosa RC, Ferreira SRS. Antioxidant and antimicrobial activities of shiitake (*Lentinula edodes*) extracts obtained by organic solvents and supercritical fluids. *Journal of Food Engineering* (2007) 80(2): 631-638.

Koh K. An approach for integrated production of mushrooms and ruminant feed. In: *Developing Modern Livestock Production in Tropical Countries*, CRC Press, pp. 4-5, 2023.

Lai LP, Lee MT, Chen CS, Yu B, Lee TT. Effects of co-fermented *Pleurotus eryngii* stalk residues and soybean hulls by *Aureobasidium pullulans* on performance and intestinal morphology in broiler chickens. *Poultry Science* (2015) 94(12): 2959-2969.

Lee TT, Ciou JY, Chiang CJ, Chao YP, Yu B. Effect of *Pleurotus eryngii* stalk residue on the oxidative status and meat quality of broiler chickens. *Journal of Agricultural and Food Chemistry* (2012) 60(44): 11157-11163.

Lee S, Im JT, Kim S, Kim Y, Kim M, Lee J, Lee H. Effects of dietary fermented *Flammulina velutipes* mycelium on performance and egg quality in laying hens. *International Journal of Poultry Science* (2014) 13(11): 637-644.

Lee MT, Lin WC, Wang SY, Lin LJ, Yu B, Lee TT. Evaluation of potential antioxidant and anti-inflammatory effects of *Antrodia cinnamomea* powder and the underlying molecular mechanisms via Nrf2-and NF-κB-dominated pathways in broiler chickens. *Poultry Science* (2018) 97(7): 2419-2434.

Lee MT, Lin WC, Lin LJ, Wang SY, Chang SC, Lee TT. Effects of dietary *Antrodia cinnamomea* fermented product supplementation on antioxidation, anti-inflammation, and lipid metabolism in broiler chickens. *Asian-Australasian Journal of Animal Sciences* (2020) 33(7): 1113.

Li S, Shah NP. Effects of various heat treatments on phenolic profiles and antioxidant activities of *Pleurotus eryngii* extracts. *Journal of Food Science* (2013) 78(8): C1122-C1129.

Liu Y, Zhao C, Lin D, Lan H, Lin Z. Effects of *Ganoderma lucidum* spent mushroom substrate extract on milk and serum immunoglobulin levels and serum antioxidant capacity of dairy cows. *Tropical Journal of Pharmaceutical Research* (2015a) 14(6): 1049-1055.

Liu Y, Zhao C, Lin D, Lin H, Lin Z. Effect of water extract from spent mushroom substrate after *Ganoderma balabacense* cultivation by using

JUNCAO technique on production performance and hematology parameters of dairy cows. *Animal Science Journal* (2015) 86(9): 855-862.

Long Y, Xiao W, Yuan C, Paengkoum P. Effects of *Flammulina velutipes* mushroom residues on growth performance, apparent digestibility, serum biochemical indicators, rumen fermentation and microbial of Guizhou black goat. *Frontiers in Microbiology* (2024) 15: 1347853.

Machado AMB, Dias ES, Santos ÉCD, Freitas RTFD. Composto exaurido do cogumelo *Agaricus blazei* na dieta de frangos de corte. *Revista Brasileira de Zootecnia* (2007) 36: 1113-1118.

Mahfuz SU, Chen M, Zhou JS, Wang S, Wei J, Liu Z, Song, H. Evaluation of golden needle mushroom (*Flammulina velutipes*) stem waste on pullet performance and immune response. *South African Journal of Animal Science* (2018) 48(3): 563-571.

Mahfuz S, He T, Liu S, Wu D, Long S, Piao X. Dietary inclusion of mushroom (*Flammulina velutipes*) stem waste on growth performance, antibody response, immune status, and serum cholesterol in broiler chickens. *Animals* (2019) 9(9): 692.

Mahfuz S, He T, Ma J, Liu H, Long S, Shang Q, Zhang L, Yin J, Piao X. Mushroom (*Flammulina velutipes*) stem residue on growth performance, meat quality, antioxidant status and lipid metabolism of broilers. *Italian Journal of Animal Science* (2020) 19(1): 803-812.

Mandale NR, Deshpande KY, Thakare SO, Morkhade SJ, Panchbhai GJ, Deshmukh SG Effect of dietary inclusion of mushroom waste on growth, nutrient digestibility, blood metabolites and economics in growing Berari goats. *Animal Nutrition and Feed Technology* (2023) 23(2): 261-277.

Manzi P, Pizzoferrato L. Beta-glucans in edible mushrooms. Food Chemistry (2000) 68(3): 315-318.

Martín C, Zervakis GI, Xiong S, Koutrotsios Strætkvern KO. Spent substrate from mushroom cultivation: exploitation potential toward various applications and value-added products. *Bioengineered* (2023) 14(1): 2252138.

Martínez Ramírez O, Bermúdez Savón RC, Rodríguez Bertot R, García Oduardo N. Comportamiento productivo de conejos alimentados con dietas que incluyen sustrato remanente de la producción de setas. *Revista de Producción Animal* (2018) 30(2): 25-31.

Masri HJ, Maftoun P, Malek RA, Boumehira AZ, Pareek A, Hanapi SZ, Enshasy HE. The edible mushroom *Pleurotus spp.*: II. Medicinal values. *International Journal of Biotechnology for Wellness Industries* (2017) 6(1): 1-11.

Mau JL, Chao GR, Wu, KT. Antioxidant properties of methanolic extracts from several ear mushrooms. *Journal of Agricultural and Food Chemistry* (2001) 49(11): 5461-5467.

Mohd Hanafi FH, Rezania S, Mat Taib S, Md Din MF, Yamauchi M, Sakamoto M, Hara H, Park J, Ebrahim SS. Environmentally sustainable applications of agro-based spent mushroom substrate (SMS): an overview. *Journal of Material Cycles and Waste Management* (2018) 20: 1383-1396.

Morshedy SA, Gad KM, Basyony MM, Zahran SM, Ahmed MH. The feasibility of partial replacement of berseem hay by spent mushroom (*Pleurotus ostreatus*) substrate in rabbit diets on growth performance, digestibility, caecum fermentation, and economic efficiency. *Archives of Animal Nutrition* (2023) 77(6): 421-436.

Mthana MS, Mthiyane DMN. Low dietary oyster mushroom spent substrate limitedly ameliorates detrimental effects of feeding combined marula seed cake and mucuna seed meal as soya bean replacements in broiler chickens. *Tropical Animal Health and Production* (2024) 56(1): 37.

Nan X, Qing-Jiu T, Jin-Song Z. Analysis of recycled nutritive com-ponents in edible mushroom by-products. *Microbiology China* (2015) 42:1929–1935

Noopan P, Chanjula P, Wattanasit S. Effects of using spent mushroom (*Cordyceps militaris*) substrate on growth performance and blood metabolite of broilers. *Kaen Kaset Khon Kaen Agriculture Journal* (2019) 47(Suppl. 1): 423-428.

Park JH, Kim SW, Do YJ, Kim H, Ko YG, Yang BS, Shin D, Cho YM. Spent mushroom substrate influences elk (*Cervus elaphus canadensis*) hematological and serum biochemical parameters. *Asian-Australasian Journal of Animal Sciences* (2012) 25(3): 320-324.

Parola S, Chiodaroli L, Orlandi V, Vannin C, Panno L. *Lentinula edodes* and *Pleurotus ostreatus*: functional food with antioxidant-antimicrobial activity and an important source of Vitamin D and medicinal compounds. *Functional Foods in Health and Disease* (2017) 7(10): 773-794.

Qi Q, Peng Q, Tang M, Chen D, Zhang H. Microbiome analysis investigating the impacts of fermented spent mushroom substrates on the composition of microbiota in weaned piglets Hindgut. *Frontiers in Veterinary Science* (2020) 7: 584243.

Rangubhet KT, Mangwe MC, Mlambo V, Fan YK, Chiang HI. Enteric methane emissions and protozoa populations in Holstein steers fed spent mushroom (*Flammulina velutipes*) substrate silage-based diets. *Animal Feed Science and Technology* (2017) 234: 78-87.

Reis FS, Barros L, Martins A, Ferreira IC. Chemical composition and nutritional value of the most widely appreciated cultivated mushrooms: An interspecies comparative study. *Food and Chemical Toxicology* (2012) 50(2): 191-197.

Ritota M, Manzi P. Pleurotus spp. cultivation on different agri-food by-products: Example of biotechnological application. *Sustainability* (2019) 11(18): 5049.

Saavedra-Castillo DM, Martínez MA, Piloni-Martini J, Quintero-Lira A, Soto-Simental S. Enzymes from *Pleurotus ostreatus* spent substrate used for fattening rabbits: effects on productive performance, carcass traits and meat characteristics. *Chilean Journal of Agricultural & Animal Sciences* (2023) 39(3): 351-358.

Seok JS, Kim YI, Lee YH, Choi DY, Kwak WS. Effect of feeding a by-product feed-based silage on nutrients intake, apparent digestibility, and nitrogen balance in sheep. *Journal of Animal Science and Technology* (2016) 58(1): 1-5.

Simitzis PE, Deligeorgis SG. Agroindustrial by-products and animal products: a great alternative for improving food-quality characteristics and preserving human health. In: *Food quality: balancing health and disease*. Academic Press, pp. 253-290, 2018.

Song YM, Lee SD, Chowdappa R, Kim HY, Jin SK, Kim IS. Effects of fermented oyster mushroom (*Pleurotus ostreatus*) by-product supplementation on growth performance, blood parameters and meat quality in finishing Berkshire pigs. *Animal* (2007) 1(2): 301-307.

Stefan RI, Vamanu E, Angelescu GC. Antioxidant activity of crude methanolic extracts from *Pleurotus ostreatus*. *Research Journal of Phytochemistry* (2015) 9(1): 25-32.

Stojković D, Reis FS, Glamočlija J, Ćirić A, Barros L, Van Griensven LJ, Ferreira ICFR, Soković M. Cultivated strains of *Agaricus bisporus* and *A. brasiliensis*: chemical characterization and evaluation of antioxidant and antimicrobial properties for the final healthy product–natural preservatives in yoghurt. *Food & Function* (2014) 5(7): 1602-1612.

Taofiq O, Calhelha RC, Heleno S, Barros L, Martins A, Santos-Buelga C, Queiroz MJ & Ferreira IC. The contribution of phenolic acids to the anti-inflammatory activity of mushrooms: Screening in phenolic extracts, individual parent molecules and synthesized glucuronated and methylated derivatives. *Food Research International* (2015) 76: 821-827.

Toghyani M, Tohidi M, Gheisari A, Tabeidian A, Toghyani M. Evaluation of oyster mushroom (*Pleurotus ostreatus*) as a biological growth promoter on performance, humoral immunity, and blood characteristics of broiler chicks. *The Journal of Poultry Science* (2012) 49(3): 183-190.

Wang CL, Chiang CJ, Chao YP, Yu B, Lee TT. Effect of *Cordyceps militaris* waster medium on production performance, egg traits and egg yolk cholesterol of laying hens. *The Journal of Poultry Science* (2015) 52(3): 188-196.

Yang JH, Lin HC, Mau JL. Antioxidant properties of several commercial mushrooms. *Food Chemistry* (2002) 77(2): 229-235.

Yuan C, Wu M, Chen X, Li C, Zhang A, Lu W. Growth Performance and Hematological Changes in Growing Sika Deers Fed with Spent Mushroom Substrate of *Pleurotus ostreatus*. *Animals* (2022a) 12(6): 765.

Yuan C, Wu M, Tahir SM, Chen X, Li C, Zhang A, Lu W. Velvet Antler Production and Hematological Changes in Male Sika Deers Fed with Spent Mushroom Substrate. *Animals* (2022b) 12(13): 1689.

Zaboli K, Kalvandi S, Malecky M, Nasrabadi M. Nutritional Value of Spent Mushroom (*Agaricus bisporus*) Compost Silage Treated with Different Level of Molasses in Sheep Feeding. *Iranian Journal of Applied Animal Science* (2023) 13(1): 57-65.

Zakil FA, Xuan LH, Zaman N, Alan NI, Salahutheen NAA, Sueb MSM, Isha R. Growth performance and mineral analysis of *Pleurotus ostreatus* from various agricultural wastes mixed with rubber tree sawdust in Malaysia. *Bioresource Technology Reports* (2022) 17: 100873.

Zhang B, Li Y, Zhang F, Linhardt RJ, Zeng G, Zhang A. Extraction, structure and bioactivities of the polysaccharides from *Pleurotus eryngii*: A review. *International Journal of Biological Macromolecules* (2020) 150: 1342-1347.

Zisopoulos FK, Ramírez HAB, van der Goot AJ, Boom RM. A resource efficiency assessment of the industrial mushroom production chain: The influence of data variability. *Journal of Cleaner Production* (2016) 126: 394-408.

Chapter 3

Eggshell Waste: A Strategy for Food Fortification

Luz Marina Gómez-Alvarez, PhD and José E. Zapata*, PhD

Research Group on Nutrition and Food Technology, NUTEC, University of Antioquia, Medellín, Colombia

Abstract

Chicken eggshells (*Gallus gallus domesticus*) are currently undervalued or discarded as waste from the agri-food industry, often without any attempt at recovery, causing environmental impact. Nevertheless, this represents an economic opportunity for producers, as these wastes contain high levels of calcium carbonate ($CaCO_3$) and matrix protein components, with various potential applications across different productive sectors. Among the alternatives for utilizing this byproduct in the food industry are technological applications, calcium supplements, or ingredients for food fortification. However, due to its poor solubility and absorption in the body, $CaCO_3$ must be transformed to add value to this biomaterial. Therefore, as a recovery and valorization strategy for eggshells, this chapter reviews topics such as the structure and composition of the eggshell, the recovery of eggshells as a source of $CaCO_3$, and the mechanisms for obtaining $CaCO_3$ nanoparticles (NPs) through nanotechnological processes, as well as the characterization of the NPs obtained from this biomaterial, along with its uses and applications.

---

* Corresponding Author's E-mail: edgar.zapata@udea.edu.co.

In: Agro-Industrial Wastes
Editor: Peter Clements
ISBN: 979-8-89113-688-5

Keywords: eggshell, $CaCO_3$, nanoparticles, top-down, fortification

Introduction

Currently, several governments are actively promoting the concept of the 'circular economy', which aims to boost sustainable development (Quina et al., 2017). Like the sustainability perspective, the 'circular economy' addresses three key issues: environmental impact, resource scarcity, and economic benefits (Lieder and Rashid, 2016). In practice, the food industry faces significant environmental challenges due to the management of large quantities of waste that generate pollution. This makes it imperative to develop optimized systems for the treatment of these wastes, to the extent that their utilization has become a priority for sustainable development (Martin-Luengo et al., 2011).

Calcium carbonate ($CaCO_3$) is an essential raw material in the food industry (Fadia et al., 2021). It is an insoluble, inorganic salt with a high calcium content (40%) and is used as an additive to induce technological changes (Nawaz et al., 2020). $CaCO_3$ has multiple functions, such as neutralization to correct excessive acidity, improvement of firmness or texture as a hardener, prevention of clumping or wetting in dehydrated vegetables, its use in baking powders in the baking sector, as well as its ability to cross-link with gluten proteins in weak flours, thereby increasing dough stiffness. Additionally, it is utilized in food fortification as a source of calcium (Quddoos et al., 2022). However, in this context, it is important to consider the bioaccessibility and bioavailability of $CaCO_3$, as its limited solubility, possibly associated with the currently used particle size (10 to 12 μm), restricts its application in food processing (Zhang et al., 2021).

There is considerable interest in the search for new sources of calcium carbonate, as calcium carbonate obtained from non-renewable sources such as limestone rocks and its synthetic production have limitations in terms of bioavailability. Although research has been carried out to obtain $CaCO_3$ from bone meal and oyster shells, traces of metals such as lead, aluminum, cadmium, and mercury have been identified, which carry a high risk of toxicity (Murakami et al., 2007). In contrast, materials such as eggshells are considered a promising alternative to produce $CaCO_3$ without contaminants. However, eggshell waste is discarded in large quantities without any pre-treatment, despite its numerous potential applications. Therefore, in recent years, research has been conducted to transform this waste into a valuable product (Hamideh and Akbar, 2018).

Eggshell is a biomaterial with a unique structure, mainly composed of a mineral phase and interstitial membranes. Its chemical composition reveals that the main component is calcium carbonate (94 - 95%) in the form of calcite, 3% organic material corresponding to the protein structure of the membrane, while the residual material contains magnesium carbonate and calcium phosphate (Baláž, 2018). This resource is utilized for a wide range of purposes, spanning from fertilizer and animal feed processes to processing for human consumption. Its most prominent application is as a source of calcium in the preparation of bioceramics (hydroxyapatite and tricalcium phosphate), composite and semiconductor materials, as well as drug delivery agents (Guru and Dash, 2014). Additionally, it serves as a dehalogenation agent, sorbent for copper removal and recovery, coating pigments for paper, adsorbent for heavy metals and organic dyes (Oliveira et al., 2013), and as a biofilter to improve the properties of polymer nanocomposites, among other uses (Hassan et al., 2013).

The transformation of this biomaterial can be achieved through processes employing top-down synthesis approaches or methods, with one of the most commonly employed techniques being mechanochemistry, which involves the use of solid-state tools and mechanical energy input for various purposes (Baláž et al., 2013; Waheed et al., 2019). Eggshell powder is typically produced by grinding using various types of mills and sieves (Hassan et al., 2015; Baláž, 2018; Huang et al., 2020; Aditya et al., 2021). Consequently, mechanical attrition has been widely utilized to synthesize nanostructured materials through high-energy ball milling of single or multi-component powders. However, a drawback of producing nanomaterials via ball milling is particle agglomeration and material contamination (Huang et al., 2020). In addition to milling, other methods for preparing nanoparticles from the top down include high-pressure homogenization and ultrasonic emulsification (Jagtiani, 2022). Nevertheless, techniques used in nanomaterial synthesis still encounter several limiting factors, such as the requirement for high temperature and pressure, toxic precursors or additives, and long reaction times. In this context, the development of simple and cost-effective green synthesis methods like ultrasonic irradiation is crucial to facilitate their applicability not only in laboratory and pilot-scale studies but also in industry (Kamali et al., 2021). Therefore, the inherent advantages of sonochemistry can accelerate the commercialization of nanotechnology across various application areas, to the point where sonochemical processing is regarded as one of the most efficient techniques for generating new materials with remarkable properties (Hassan et al., 2013).

The Hen's Egg Shell (*Gallus gallus domesticus*)

Structure, Formation, and Composition of the Chicken Eggshell

The hen eggshell (*Gallus gallus domesticus*) is a complex biomineral structure with a perfectly defined polycrystalline organization along the calcified eggshell (Wilson, 2017). It serves to protect the egg contents from microbial and physical environments, providing calcium for the growing embryo (Nys et al., 2004; Gonzalez, 2020) and allowing water and gas exchange between the external environment and the embryo. This capacity is facilitated by the protein composition of the cuticle (Wellman-Labadie et al., 2008).

The avian eggshell consists of six distinct layers. The two inner layers are non-calcified membranes, each comprising a network of fibers surrounding the albumen (Rivera et al., 2022). The inner zone of the calcified shell consists of irregular cones corresponding to the mammillary protrusion layer, with the tips of these cones penetrated by fibers of the outer membrane (Wellman-Labadie et al., 2008). Extending beyond the bases of the cones is the palisade layer, which culminates in a thin vertical crystal layer where crystals are aligned perpendicular to the shell surface (Hincke et al., 2000; Shang et al., 2022). The outer layer, known as the cuticle, is an organic layer deposited on the egg surface. This layer contains a thin film of hydroxyapatite crystals in its inner zone (Stapane et al., 2020), with approximately two-thirds corresponding to surface eggshell pigments (Nys et al., 2004), which are related to the egg's lysozyme and ovotransferrin concentration (Lewko et al., 2021). Additionally, eggshells possess between 7,000 and 17,000 funnel-shaped pore channels that facilitate water and gas exchange (Shang et al., 2022).

Eggshell formation occurs through a process of calcification triggered by precipitation on the shell membrane during the transit of the egg through the oviduct (Nys, 2001). This process, highly controlled in temporal and spatial terms, unfolds as the yolk progresses through the oviduct, gradually acquiring the white in the magnum and the shell membranes in the isthmus. At the distal red isthmus, organic aggregates called mammillary protrusions form, initiating calcium carbonate nucleation (Nys et al., 2004). Subsequently, in the next phase of shell formation, the egg acquires its final shape in the uterus by filling it with albumin and soluble organic precursors (Nys et al., 1991; Gautron et al., 1997). During this process, the egg rotates as calcium carbonate's mammillary and palisade layers are deposited. Mineralization stops approximately one and a half hours before oviposition, at which time the organic cuticle is deposited (Zhou et al., 2011; Sah and Mishra, 2018). Mineralization is thought to halt due to inhibition by a specific component of the uterine fluid, as the medium remains supersaturated with calcite at this point (Nys et al., 2004). This suggests precise regulation of shell mineralization during shell formation.

The eggshell represents 9 to 12% of the egg weight. Most of the calcified zone is mainly composed of calcium carbonate ($CaCO_3$, 95%), in the form of calcite, its most stable polymorph. Additionally, magnesium carbonate ($MgCO_3$, 1%) and calcium phosphate ($Ca_3(PO_4)_2$, 1%) are present (Da Silveira Pinto and De Souza, 2019; Ajala et al., 2018; Waheed et al., 2020). The remaining 3% consists of an organic matrix composed mainly of fibrillar proteins with disulfide cross-links and type I, V, and X collagen in the eggshell membranes, and of proteoglycans and glycoproteins in the calcified layers (Leach, 1982; Nys et al., 1991; Nys et al., 2004; Hincke et al., 2012). The mechanical properties of the shell are attributed to interactions between mineral components and organic matrix proteins. Proteins such as podocalyxin-32, ovocalyxin-36, ovocleidin-116, and osteopontin have been identified as key elements in eggshell mineralization, together with clusterin, lysozyme, ovotransferrin, and collagen, which are involved in the regulation of calcium carbonate crystallization and its vesicular transport (Aina et al., 2022; Le Roy et al., 2021; Shang et al., 2022).

Uses and Applications of Eggshells

Eggshells are widely recognized as one of the most abundant natural bio-wastes and have found numerous applications in various fields (Shang et al., 2022). With their rich content of bioactive compounds, there is a growing interest within the scientific community to explore and develop eggshell-derived products for technological and commercial use (Cordeiro and Hincke, 2011). One promising avenue for eggshell utilization is in the steel industry. In 2019, over 50% of the calcium carbonate used in this industry was sourced from limestone, presenting an opportunity for eggshell waste to serve as an alternative material. Additionally, limestone consumption is significant in the building and construction sector, with the remainder allocated across industries such as water treatment, agriculture, paper, plastics, and paints (Lewicka et al., 2020; Market, 2020).

Calcium compounds, extracted from sources such as eggshells, find diverse applications in the biomedical, nutritional, and pharmaceutical sectors. Products like calcium citrate and calcium phosphate derived from these sources are projected to reach a market value of USD 900 million by 2025 (Ahmed et al., 2021). While calcium supplements are commonly sourced from dried shrimp, fish, oyster shells, corals, and seaweed, eggshells have emerged as an efficient and acceptable source for both animal feed and human health, surpassing oyster shell waste in calcium extraction (Schaafsma et al., 2000; Lewicka et al., 2020; Ahmed et al., 2021). Several calcium supplements derived from eggshells are available on the commercial market. For instance, Bone Health Original™ capsules, which prioritize bone health and advocate eggshells as a superior source of easily absorbable calcium. Another product, EGGNOVO's OVOCET®, utilizes eggshells as a highly absorbable calcium source for both humans and animals. Additionally, IrRAWsistible offers the Eggshell Calcium Supplement™, tailored for the pet industry (Ahmed et al., 2021).

Research on intellectual property applications of eggshells has resulted in approximately 673 patents, with 80% of these issued in the last two decades, mostly focused on biotechnological applications (Ahmed et al., 2021). These applications include biomedical uses, such as the manufacture of wound and skin ulcer dressings (Shagdar and Myagmar, 2019), its use as a scaffolding component for bone regeneration (Wu et al., 2019), and its utilization as a raw material in the production of antimicrobial agents (Shang et al., 2022). Chelated calcium obtained from eggshells is utilized in skincare products, which have whitening, exfoliating, and moisturizing properties (Ahmed et al., 2021).

In the food industry, eggshells find a wide array of applications, offering additional nutritional support through supplementation or as an additive in curing processes, flavoring, and antimicrobial roles, alongside its stabilizing properties (Aditya et al., 2021; Ahmed et al., 2021). Eggshell powder serves as a rich source of "biological" calcium for oral dietary supplements aimed at enhancing bone density and addressing age-related bone loss (Waheed et al., 2020; Xiao et al., 2021). Moreover, it has been utilized in animal feed, where it is incorporated into the diets of laying hens to enhance bird health, boost egg production, and maintain high shell quality (Ahmed et al., 2021).

Eggshells have a variety of uses. They can serve as plant growth promoters by converting $CaCO_3$ into more soluble and efficient soil fertilizers (Baláž et al., 2021). Additionally, they can be transformed into calcium oxide, a natural ingredient in soaps that mitigates concerns about human skin toxicity associated with conventional chemical compounds (Ahmed et al., 2021). Furthermore, these shells act as a source of soluble proteins with valuable bioactive properties (Hincke et al., 2012). Other applications include biodiesel oil catalysis, the production of biodegradable plastics and synthetic paper, and their use as a natural biosorbent for heavy metals, dyes, and organic compounds in wastewater treatment and agricultural remediation processes (Kavitha et al., 2019; Kasmuri and Zait, 2018; Zhou et al., 2019; Nuhu et al., 2012; Baláž, 2014; Ahmed et al., 2019; Guijarro-Aldaco et al., 2011; Sabir et al., 2021).

Valorization and Up-Scaling of Eggshell Waste

Applications leveraging eggshell waste have seen significant growth in recent years, driving not only research, patents, and new product development (Ahmed et al., 2021) but also the establishment of technology-based companies that capitalize on this waste and its bioactive compounds. An exponential increase in patent filings has been observed over the last two decades. Eggshell waste holds substantial market potential due to its calcium carbonate ($CaCO_3$) and matrix protein content. The global market for calcium carbonate, encompassing both ground and precipitated forms, is projected to reach USD 28.98 billion by the end of 2025, with a compound annual growth rate of 5.78% (Ahmed et al., 2021). This suggests that eggshell residues could serve as partial substitutes for commercial products like limestone, pure calcium carbonate, soil conditioners, biomaterials, food additives, or supplements, as well as finding applications in cosmetics or pharmaceuticals, among others (Waheed et al., 2019; Mignardi et al., 2020).

The European Commission and the Seventh Framework Programme (FP7) funded the project "Separating ES and its membrane to turn ES waste into valuable source materials" in 2012 to address the environmental issues arising from eggshell (ES) waste. A budget of 1.5 million euros was allocated for this purpose (Ahmed et al., 2021; Platon et al., 2022).

Consequently, a prototype capable of processing 60 kg/h of waste, equivalent to 10,000 eggs/hour, was developed, achieving high-quality eggshell and membrane separation (Ahmed et al., 2021). Similarly, in Canada, Cordeiro and Hincke (2011) reported similar projects funded for the recovery of proteins from eggshell waste for medical applications and health product development, as well as obtaining eggshell powder as a functional food ingredient.

EGGNOVO, a Spanish company, stands as a successful model for valorizing eggshell waste. It specializes in offering branded functional ingredients derived from eggshells across various sectors including food, cosmetics, nutraceuticals, and pharmaceuticals (Aguirre et al., 2017; Andres et al., 2018; Gil-Quintana and Molero, 2020). This achievement stemmed from a project funded by the European ECO-INNOVATION initiative, which facilitated the development of commercial prototypes for four eggshell separation systems: ovocet®, ovomet®, ovopet®, and ovoderm®. For instance, Ovocet has been validated for application in plastic matrices and food products, presenting an eco-friendly substitute to conventional calcium carbonate. Ovomet underwent preclinical and clinical evaluations to assess its effectiveness in addressing joint pathologies. Ovopet was integrated as a feed ingredient, with specific research conducted on dogs with hip dysplasia. Lastly, Ovoderm concentrates on enhancing skin hydration, elasticity, and regeneration (Eggnovo, 2016).

Instances of industrial scale-up and the utilization of eggshell waste exemplify the potential of this material for various applications across several sectors. This transformation of bio-waste into value-added products benefits both the environment and the economic prospects of producers.

Calcium Carbonate

Calcium carbonate ($CaCO_3$) is a chemical compound composed of carbon, oxygen, and calcium. It forms when calcium and carbonate ions combine in an aqueous solution, as demonstrated in the chemical reaction: $H_2O + CO_2 + CaCO_3 \leftrightarrow Ca^{2+} + 2HCO^{3-}$ (Moriconi et al., 2022). This white, tasteless mineral is naturally occurring in materials such as chalk, limestone, and marble. Its quality and applications depend on factors such as purity, whiteness, homogeneity, thickness, and particle size (Fu et al., 2022). Recognized as safe by the Food and Drug Administration (GRAS), $CaCO_3$ has attracted significant research interest in recent years due to its potential.

In addition to being extracted from rocks like limestone, it is a primary component of shells from marine organisms, snails, charcoal balls, pearls, and eggshells, making it an abundant biomineral in both organisms and nature (Al Omari et al., 2016; Chen et al., 2019; Stoica-Guzun et al., 2012; Zhang et al., 2013; Fernandes et al., 2014).

The two main commercial forms of calcium carbonate are GCC and PCC. GCC is obtained by finely grinding or micronizing highly pure limestones, typically with a $CaCO_3$ content of over 98.5%. In the United States, it is defined as a product derived from limestone or dolomite with a minimum purity of 97% and a grain size of less than 45 mm. In Europe, the definition excludes products from dolomites, utilizing limestone, marble, or chalk as raw materials (Krammer et al., 2002; Das et al., 2020). PCC is purer than the original carbonate rock and contains less silica, magnesium, and lead. Its morphology and size differ from GCC, as GCC has an irregular rhombohedral shape under magnification, while PCC has uniform and regular particles. Manufacturing PCC with defined morphology and particle size is crucial due to its numerous applications in various industries (Xiang et al., 2002). The complexity of its structural isomerism and crystalline morphology has motivated ongoing research on calcium carbonate (Jimoh et al., 2018).

Sources and Types of $CaCO_3$

Calcium carbonate exists in various polymorphs with different crystal structures. These include three anhydrous polymorphs: vaterite, aragonite, and calcite, which are named in increasing order of thermodynamic stability (Wang et al., 2012; Cölfen, 2003). Additionally, there are two hydrated polymorphs: monohydrocalcite ($CaCO_3 \cdot H_2O$) and ikaite ($CaCO_3 \cdot 6H_2O$), along with several amorphous phases (Dhami et al., 2013). Calcite is the most common polymorph, especially in rocks, due to its higher stability compared to aragonite and vaterite. Studies have shown that aragonite can transform into calcite at temperatures around 400-450°C during heat treatment (Chervonnyi et al., 2003; Park et al., 2021). Over geological time, vaterite and aragonite are converted to calcite, resulting in $CaCO_3$ derived from limestone predominantly consisting of calcite.

Table 1. $CaCO_3$ applications in different industrial sectors

Industrial Sector	Uses/Applications of $CaCO_3$
Paper	Calcium carbonate serves as a filler in cellulose pulp, enhancing the paper's whiteness and strength. It has supplanted kaolin and talc in the papermaking process by enabling a neutral and alkaline approach. Acting as both a filler and coating, it enhances several paper attributes including brightness, opacity, satin finish, porosity, and printability (Zhou et al., 2019; Zárybnická et al., 2022).
Paints	$CaCO_3$ is widely used due to its excellent hiding power, stability, and dispersibility. In addition, it serves as an economical alternative to titanium dioxide and aluminum silicate as extenders in paints (Karakaş et al., 2015). Calcite improves various properties such as spreadability, opacity, matt appearance, and coloring effect. It also adjusts gloss, improves adhesion, increases solids content, reduces costs, enhances whiteness and weatherability, and improves abrasion resistance (Ersoy et al., 2021).
Plastics and polymers	In the extrusion process, calcium carbonate is added to the product before extrusion to improve some properties that make the process more efficient; it increases the thermal conductivity of the product so that extrusion can be performed faster, improves homogeneity and dispersion with a consequent improvement in the condition of the extruder (Cha et al., 2004).
Ceramics	Calcium carbonate is widely used in ceramics because of its whiteness and purity, as well as its low humidity and good machinability (Vilarinho et al., 2022).
Chemical industry	Calcium carbonate is used in a variety of applications, such as gas desulphation and lime oxide production (Rohim et al., 2014; Aina et al., 2022). It is also used as a mineral filler in the manufacture of detergent soaps to improve moisture retention, consistency, and drying of the final mass (Abdullah et al., 2022).
Cosmetics	It is mainly used for its mechanical properties, it is suitable for use in toothpaste as an abrasive due to its low metal content, it is also used in make-up powders to add volume to the product (Oliveira et al., 2013).
Glass	$CaCO_3$ is used as a filler, improves the mechanical properties of the glass, and confers higher strength (Chen et al., 2004; Muhammad et al., 2022).
Agriculture and Livestock	Calcium carbonate is used for various purposes. Mainly, its basic character is used to raise the pH of acid soils, thus achieving a neutral level suitable for the cultivation of fruit and vegetables. In addition, it provides minerals such as calcium to the soil, or magnesium in the case of dolomites. This increase in soil pH leads to a reduction in aluminum concentration, increased phosphorus availability, and improved nitrogen fixation, as well as increased nitrification and mineralization (Nyamaizi et al., 2022). Additionally, it is used as a calcium supplement for animals (Summa et al., 2022).
Human food	$CaCO_3$ is used in human food due to its various physical, mechanical, and chemical properties. It is used in dairy products to reduce acidity, increase whiteness, and enrich calcium content. It is also used for this purpose in beverages such as water, soy milk, or rice milk. In addition, it is used to harden fruits and vegetables before freezing or canning, as a drying agent in dehydrated products, surface coloring in chewing gum, to prevent caking of powdered foods, and in the bread industry to improve fermentation and increase volume (Jimoh et al., 2018; Waheed et al., 2019; Waheed et al., 2020; Vandeginste, 2021, Platon et al., 2022).
Health/ Medicine	Biocompatible calcium carbonate dissociates at low pH, making it a compound with interesting potential in the field of biomedicine, which is why it has been used to develop drug delivery systems via subcutaneous, oral, intramuscular, or intravenous routes (Fadia et al., 2021; Fu et al., 2022). pH-sensitive drug delivery systems have been developed for cancer treatment using drug impregnation or encapsulation techniques (Shen et al., 2017).

Calcite finds extensive applications in the cement and metallurgy industries (Yang and Xu, 2011), whereas aragonite is frequently utilized as a filler in biomedical materials and novel composites (Santos et al., 2012). Despite being the least common polymorph, vaterite stands out for its potential biomedical applications due to its high specific surface area, solubility in water and dispersion, and lower density compared to other crystalline polymorphs (Fu et al., 2013). Porous vaterite particles have demonstrated potential for controlled drug release (Parakhonskiy et al., 2011), and vaterite nanoparticles have been investigated for the treatment of degenerative bone defects and diseases (Schröder et al., 2015). Therefore, the production of vaterite with various surface structures and properties holds significant importance (Polat and Sayan, 2020). The formation of any of these polymorphs depends on parameters such as temperature, supersaturation, and pH of the reaction solution (Ibrahim et al., 2014). Amorphous calcium carbonate (ACC) is a metastable polymorph of $CaCO_3$ that can facilitate the biomineralization of calcite and aragonite due to its high ion concentration (Stapane et al., 2020). Although more soluble and less stable than crystalline polymorphs, ACC has been identified as a precursor of other $CaCO_3$ polymorphs, following a sequence that includes hydrated ACC, anhydrous ACC, vaterite, aragonite, and finally, stable calcite (Oriols et al., 2020).

$CaCO_3$ Applications

The global market for $CaCO_3$, estimated at USD 43.89 billion, is projected to grow at an annual rate of 2.8%. Asia dominates this market, accounting for approximately 43% of the total share. Table 1 illustrates some of its applications.

Uses of $CaCO_3$ from Eggshells

Hen eggshells, rich in $CaCO_3$, are a versatile bioceramic compound with diverse industrial applications, making them valuable biomaterials (Oliveira et al., 2013; Waheed et al., 2019; Waheed et al., 2020; Vandeginste, 2021).

$CaCO_3$ powder obtained from eggshells exhibits remarkable capability in removing heavy metals from water and soil. Studies by Liao et al. (2010) and de Paula et al. (2008) have demonstrated its effectiveness in eliminating divalent metal ions such as lead, cadmium, and copper from aqueous solutions. Additionally, Ok et al. (2011) suggest the utilization of eggshells as an alternative to $CaCO_3$ for immobilizing heavy metals in soils. This material is widely employed as a filler in various applications including the pharmaceutical industry, soil stabilization, medical and dental preparations, food additives, calcium supplements, agricultural fertilizers, and bone-implant components due to its excellent dispersibility and affinity (Oliveira et al., 2013; Ahmed et al., 2021; Aina et al., 2022).

$CaCO_3$ derived from eggshells could serve as a substitute for minerals used in paper treatment, enhancing properties such as gloss, opacity, strength, appearance, and texture (Oliveira et al., 2013; Zhou et al., 2019; Tutus et al., 2020). It is also utilized as a solid catalyst in the transesterification of vegetable oils with methanol for biodiesel production (Wei et al., 2009). In orthopedics, eggshells are used as a substitute for bone grafts to treat congenital diseases, tumors, or trauma. Biodegradable support structures have been developed to improve bone regeneration (Huang et al., 2020). In dentistry, eggshell powder solution has been utilized to remineralize carious lesions of the enamel and to fabricate dental prostheses (Mony et al., 2015). Hydroxyapatite derived from eggshells is valued for its excellent biocompatibility with soft tissues such as skin, muscle, and gums, making it an ideal candidate for orthopedic and dental implants or implant components (Lee et al., 2012). In environmental applications, it is used in water and soil decontamination as a low-cost absorbent for eliminating cadmium, lead, and copper ions, thereby immobilizing heavy metals (Ahmed et al., 2021).

The food industry faces a significant challenge in waste management and technology development to optimize waste utilization, such as eggshells, not only due to their large volume but also their environmental impact. Hence, it is essential to focus on the added value of eggshells, particularly their utility as a calcium source, optimizing their extraction and presentation to enhance bioavailability. Numerous studies have explored integrating eggshells into various food products to bolster their calcium content. These include calcium-fortified pork sausages (Daengprok et al., 2002), yogurt supplemented with eggshell nanopowder (Al Mijan et al., 2014), calcium-fortified diet biscuits (Hassan, 2015), roasted coffee and calcium-fortified ground (de Paula et al., 2014), calcium-enriched chocolate cake (Ray et al., 2017), calcium-fortified bread (Platon et al., 2022), as well as homemade food products such as pizza and pasta (Brun et al., 2013), and confectionery items like snacks with high calcium content (Pokorski and Hoffmann, 2022). However, challenges in calcium absorption persist, which could potentially be addressed by transforming $CaCO_3$ at the nanoscale to leverage its bioactive mechanism. Studies suggest that the porous structure of eggshell calcium allows for better absorption in the human intestine (Aditya et al., 2021).

Nanotechnology in Food Science

Global food production and its supply chain are transitioning towards more balanced production with reduced waste, prioritizing public health and consumer needs, and addressing industry challenges (Pushparaj et al., 2022). In this context, nanotechnology has emerged as a promising strategy that has revolutionized multiple fields, including pharmaceuticals, agriculture, and nutrition (Nile et al., 2020). Nanotechnology offers the potential to be integrated into various stages of food production to enhance food quality and safety, employing tools such as sensors for pathogen and contaminant detection, as well as devices for tracking and ensuring product safety (Duncan, 2011).

In the food sector, Regulation (EU) No 2015/2283 of the European Parliament (2015) defines nanomaterials as intentionally produced materials with one or more dimensions of 100 nm or less, or composed of discrete functional parts, either internally or on the surface, many of which have dimensions on the order of 100 nm or less. This definition encompasses structures, agglomerates, or aggregates that may have sizes larger than 100 nm but retain characteristic nanoscale properties (Gómez-Llorente et al., 2022).

Nanotechnology applied in the food industry encompasses various areas, including food safety, shelf-life extension, creation of healthier products with reduced fat, sugar, and salt content, enhancement of flavor, aroma, color, and texture, as well as pathogen, toxin, and pesticide detection. Additionally, it enables the delivery of functional foods, supplements, and nutrients to enhance their bioavailability (He and Hwang, 2016; Arshad et al., 2021). The unique physical, chemical, and biological properties of nanomaterials, such as size, shape, chemical composition, stability, crystalline structure, and larger surface area, impart valuable characteristics that improve thermal stability, water solubility, and nutrient bioavailability (Nile et al., 2020; Arshad et al., 2021; Dash et al., 2022; He and Hwang, 2016). Nanomaterials serve as delivery systems to enhance the bioavailability of bioactive compounds like vitamins, iron, and calcium, utilizing nano delivery systems such as association colloids (casein micelles), lipid-based nanoencapsulation, nanoemulsions, nanoparticles, nanolaminates, and nanofibers (Kim et al., 2015; He and Hwang, 2016).

While nanomaterials offer potential benefits in the food industry, they also pose risks due to changes in physical and chemical properties at the nanoscale, which could lead to toxicity issues for humans, animals, and the environment (Dash et al., 2022; Gómez-Llorente et al., 2022). Consequently, nanotechnology is under extensive study, with numerous research efforts focused on developing applications, characterizing nanomaterials, and conducting toxicity studies. Moreover, it is crucial to investigate the mechanisms of action of nanomaterials and bioactive components under various human physiological conditions, necessitating further research based on clinical evidence to better understand their effects (Arshad et al., 2021).

The utilization of agroindustrial waste as substrates for nanomaterial production has gained widespread acceptance due to its economic, environmental, and technological benefits. These wastes, derived from horticultural, aquacultural, domestic, or animal sources, such as eggshells, bones, scales, skin, and viscera, contain bioactive compounds useful not only in the food industry but also in other sectors (Elemike et al., 2022). For instance, Santillán-Urquiza et al. (2017) fortified yogurts with calcium and zinc nanoparticles, observing improvements in physical properties such as consistency and firmness, along with greater solubility compared to conventional fortification. Additionally, studies like that of Erfanian et al. (2014) in fortified milk powder have demonstrated enhanced bioavailability when supplemented with nano $CaCO_3$, aligning with prior research indicating improved absorption through particle size reduction (Heaney, 2000; Park et al., 2021; Huang et al., 2009; Jeong et al., 2013; Lee et al., 2015; Huang et al., 2020; Quddoos et al., 2022).

Properties and Interactions of Nanoparticles

Nanoparticles are classified into two main categories based on their constituent materials: organic and inorganic. Organic nanoparticles are typically composed of proteins (e.g., zein nanoparticles and casein micelles), carbohydrates (nanochitin and nanocellulose nanogels), and phospholipids and lipids (liposomes). Inorganic nanoparticles, on the other hand, are primarily made of metals or metal salts such as gold, copper, titanium dioxide, silicon dioxide, or zinc oxide (McClements, 2020; Moradi et al., 2022). While many of these materials are used as food additives, their functional performance and safety in foods are influenced by their interactions with other ingredients present in the matrix, which can affect their aggregation, interfacial chemistry, and surface charge (Pushparaj et al., 2022; Moradi et al., 2022).

Nanoparticles used as food additives exhibit a diversity of compositions, sizes, shapes, and surface characteristics, which influence their behavior both in food matrices and in the human gastrointestinal tract (GIT) (Setyawati et al., 2020). Their functional attributes and fate in the GIT may vary depending on the nature of their environment (Moradi et al., 2022). Nanoparticles can interact with other components through various molecular interactions, including electrostatic, hydrophobic, van der Waals, steric, and hydrogen bonding interactions (Baláž et al., 2013; Zhou et al., 2019). Moreover, the properties of nanoparticles can be modified by their interactions with molecules present in the food matrix (McClements et al., 2020), and coronas can form around nanoparticles due to the adsorption of biological molecules such as proteins, carbohydrates, phospholipids, lipids, or mineral ions on their surfaces (Lee et al., 2015; McClements et al., 2016). The composition, structure, and behavior of these biocoronas depend on the type and organization of the molecules present, determined by their concentrations and relative affinities towards the particle surface (Kopac, 2021).

The adsorption of biological molecules on the surfaces of nanoparticles can alter their composition, structure, thickness, and charge, impacting their behavior within the human GIT. After ingestion, both particles and food complexes undergo different stages, from the mouth to the colon, interacting at each stage with specific gastrointestinal secretions, pH variations, mineral compositions, and mechanical forces, which influence their fate and toxicity. Furthermore, the properties of the mucus layer and epithelial cells lining the GIT act as physical barriers, also determining the fate of ingested nanoparticles (McClements et al., 2016; Moradi et al., 2022). Given the complexity of these interactions and their impact on various aspects such as performance, gastrointestinal fate, and toxicity of nanoparticles, it is imperative to deepen research to understand the fundamental mechanisms involved, especially through in vivo studies that can shed light on changes in gastrointestinal cell permeability induced by nanoparticles (Cao, 2022).

Top-Down and Bottom-up Processes

Nanotechnology offers two distinct approaches for obtaining nanostructured materials: top-down and bottom-up. The top-down approach involves the physical or chemical breakdown of large particles into smaller ones at the nanometer scale. Techniques such as mechanical attrition, exemplified by high-energy ball milling of powders, have emerged as effective and scalable methods for synthesizing nanostructured materials (Baláž, 2018). This method entails grinding bulk materials to the nanoscale and then stabilizing the resultant nanoparticles with colloidal protectants (Jagtiani, 2022). Impact and shear forces play pivotal roles in particle size reduction, especially in the production of nanoparticles for food applications (Sanguansri and Augustin, 2006). Shear, one of the primary forces employed in particle size reduction, is achieved through high shear systems like rotor-stator homogenizers, which facilitate dispersion via localized energy dissipation and intense shear speeds. In a recent study by Gómez-Álvarez et al. (2022), high shear dispersion preceded ultrasonic cavitation to obtain $CaCO_3$ nanoparticles from eggshells.

Common technologies utilized for size reduction (top-down) include dry grinding using ball mills, colloidal mills (e.g., Ultra Turrax), jet milling, and high-energy mills (Baláž et al., 2013). High-pressure homogenization and microfluidization are also employed in food processing to achieve size reduction, dispersion, or emulsion formation via optimal forces of cavitation, shear, and impact (Jagtiani, 2022). Ultrasonic technology facilitates the dissolution and precipitation of solids, resulting in particle size reduction and surface activation through intensive stirring (Sonawane et al., 2010). Nanoprecipitation, also referred to as antisolvent precipitation or solvent displacement, can effectively reduce particle size within the range of 50–300 nm (Hassan et al., 2014).

While ball milling remains the predominant method for manufacturing nanomaterials via top-down processes due to its simplicity, cost-effectiveness, and versatility across various materials, it does come with drawbacks such as surface defects, particle agglomeration, and material contamination (Jagtiani E., 2022). Ultrasound irradiation presents an alternative for reducing the size of nanoscale materials, known for its efficiency in producing materials with unique properties (Hassan et al., 2013). This technique, relying on acoustic cavitation, creates extreme conditions of temperature, pressure, and cooling rate within the irradiated solution (Shimpi et al., 2015). The combination of high shear dispersion and high-intensity ultrasound, as proposed by Gómez-Álvarez et al. (2022), has demonstrated efficacy in reducing particle size to the nanoscale. While top-down processes continue to advance, exploring innovative approaches such as nanoparticle synthesis from eggshells represents an emerging field with potential to enhance the utilization of this valuable agro-industrial waste, as suggested by Ahmed et al. (2021).

Conversely, the bottom-up approach involves the fabrication of materials from individual atoms that spontaneously self-assemble and self-regulate (Sanguansri and Augustin, 2016). Bottom-up methodologies typically result in fewer failures, greater homogeneity, and controlled variables (Jagtiani E., 2022), facilitating the production of nanomaterials with reduced defects and a more uniform composition. Additionally, this approach allows for the formation of thin films with a relatively simple structure, minimizing waste material during synthesis. Key parameters such as pH, sintering time, and temperature play pivotal roles in controlling the properties of nanoparticles synthesized via the bottom-up approach (Kumar et al., 2020).

$CaCO_3$ Nanoparticles

The synthesis of $CaCO_3$ nanoparticles has sparked significant interest due to their larger surface area and smaller particle size, making them valuable for various advanced applications. Typically produced through wet carbonation, often with the inclusion of surfactants to regulate crystal growth and the addition of pore formers for nanopore construction (Ibrahim et al., 2014), these nanoparticles find utility in both industrial and medical contexts. With unique physicochemical properties, nano sized $CaCO_3$ particles can enhance drug solubility, pharmacokinetics, and targeted delivery (Baláž et al., 2021), while also serving as CO_2 capture and storage agents, offering considerable potential in climate change mitigation (Xiang et al., 2002; Jimoh et al., 2018).

Nano calcium carbonate, an ultrafine solid material that emerged in the 1980s, possesses a broad particle size distribution (approximately 15-105 nm) and irregular shape, with an average particle size of about 40 nm. Its crystalline particles are cubic and partially fused, forming chains. The unique properties of nano $CaCO_3$, stemming from its nanoscale dimensions, include a rapid increase in surface atomic number, surface area, and energy, setting it apart from conventional particles. Typically produced via physical grinding, this material has found widespread application in various fields such as construction, polymers, papermaking, paints, inks, biochemistry, and others, owing to its distinct chemical and physical attributes (Fu, Q. et al., 2022).

Utilizing $CaCO_3$ nanoparticles derived from eggshells serves various purposes. The inherent pore structure of eggshells renders them ideal natural scaffolds for nanoparticle production, such as NP-$Ca(OH)_2$ (Yang et al., 2021). Furthermore, eggshells have been explored as templates to produce $CaCO_3$-(Ag) NPs, wherein they are pulverized and loaded with Ag ions before being heated at 400–600°C. The resulting $CaCO_3$-Ag NPs exhibit diverse applications, including catalysis, tissue engineering, coatings, and the manufacturing of antibacterial agents, pigments, and ceramics (Yang et al., 2021).

Apalangya et al. (2014) utilized straightforward mechanochemical milling to synthesize nanocomposites employing eggshell particles as templates, yielding Ag NPs, copper oxide NPs, and zinc oxide NPs. Meanwhile, Zhang et al. (2013) produced CuO/ZnO/eggshell (CZ/ES) composites using eggshells as solid substrates (Shang et al., 2022). In essence, the applications of $CaCO_3$ nanoparticles are diverse and backed by recent research. Baláž et al. (2021) provide a comprehensive overview of the latest advances in utilizing eggshells in catalysis and mechanochemistry. Moreover, advancements in enhancing calcium bioavailability through eggshell nanoparticles are noteworthy, as these particles have demonstrated higher water absorption capacity and lower zeta potential compared to microscale eggshell powder (Ahmed et al., 2019). Nonetheless, several challenges remain to be addressed from this perspective.

Characterization of $CaCO_3$ Nanoparticles

Nanomaterials, particularly those intended for applications in food or pharmaceutical systems, necessitate comprehensive characterization to elucidate the relationships between their structure and function, thereby enhancing absorption and bioavailability (Amalraj and Pius, 2015). Various properties of nanoparticles can significantly impact their behavior in biological systems (Aditya et al., 2021). Thus, the imperative to integrate multiple characterization methods is acknowledged to furnish comprehensive insights into their physicochemical attributes, encompassing agglomeration, chemical composition, crystal structure, particle size and distribution, purity, shape, surface area, and charge (Baláž P., 2013; Baláž M., 2018).

A. Scanning electron micrograph of eggshell nanopowder.

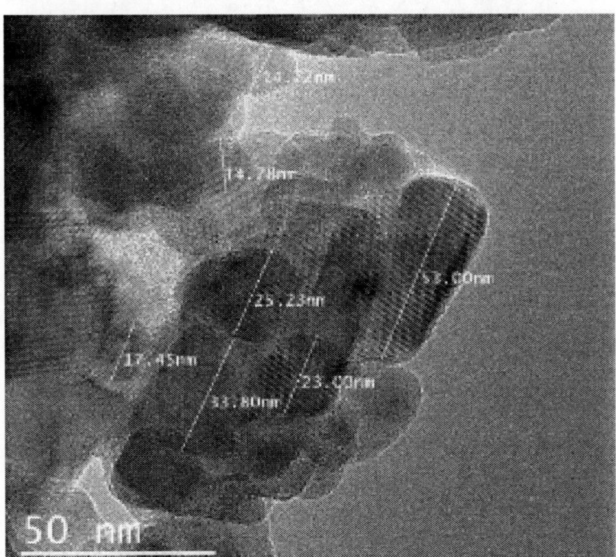

B. Transmission electron micrograph of eggshell nanopowder.
Total magnification: 145.00 kx.

Figure 1. Electron micrographs of the nanoparticles obtained from the eggshell.

Characterization methods can be categorized into two primary groups: sample preparation and separation techniques (e.g., centrifugation, nanofiltration, dialysis, chromatography), and methodologies to evaluate nanoparticles concerning their physical, chemical, or physicochemical properties. This encompasses dynamic light scattering for particle size measurement, electrophoretic mobility, polydispersity index, and zeta potential determination to assess stability and aggregation patterns (Cordeiro et al., 2011; D'Haese et al., 2013; Apalangya et al., 2014; Erfanian et al., 2017; Baláž M., 2021).

The determination of specific surface area employs techniques grounded on adsorption and desorption principles, such as the Brunauer-Emmett-Teller (BET) theorem. Atomic force microscopy (AFM) enables the assessment of surface roughness, while scanning electron microscopy (SEM) (Figure 1-A) and transmission electron microscopy (TEM) (Figure 1-B) facilitate examination of surface and internal morphology, porosity, particle, and pore size (Hassan et al., 2013, Hassan et al., 2014). Additionally, X-ray diffraction (XRD) furnishes insights into crystallinity and structure, encompassing chemical composition, crystal size, and preferred orientation (Fu et al., 2013; Gandolfi et al., 2015; Faridi and Arabhosseini, 2018; Fu J., et al., 2022; Fu K., et al., 2022).

Fortification of Foods with Calcium

In recent years, food fortification has garnered notable interest due to the growing understanding that micronutrient deficiency is a leading cause of disease worldwide (Mattar et al., 2022). The Codex Alimentarius defines food fortification as "the addition of one or more essential nutrients to a food, whether or not normally present in it, to prevent or correct demonstrated deficiencies of one or more nutrients in the food population or specific population groups" (Dwyer et al., 2015; Benkeblia, 2020). This strategy has been identified as the most cost-effective for addressing micronutrient deficiencies and has the advantage of not requiring modifications in dietary habits, potentially increasing its acceptability compared to other strategies (Bourassa et al., 2022).

Dietary calcium deficiency is a global problem, with few countries achieving adequate levels of intake of this essential mineral. The majority of the world's population consumes less than 1000 mg of calcium per day. In regions such as East, South, and Southeast Asia, as well as Nepal, average consumption is especially low, with figures below 400-500 mg/day, reaching a minimum of 175 mg/day in Nepal. In Africa and South America, average intake levels range between 400-700 mg/day, with limited access to dairy products being one of the main reasons (Galvan-Ruiz et al., 2017). On the other hand, in northern European countries, a daily calcium intake of more than 1000 mg is recorded, with Iceland standing out with an average supply of 1233 mg/day (Wawrzyniak and Suliburska, 2021; Bourassa et al., 2022).

The choice of calcium source for food or beverage formulation depends on several factors, such as its ease of use in the specific application, its effects on flavor, and the stability of the final product. In addition, the total cost and the health properties sought also influence this decision. As for health properties, it has generally been observed that the body absorbs calcium more effectively when consumed in smaller amounts spread throughout the day. Therefore, the addition of calcium is probably most effective at fortification levels up to 30% of the Daily Value (DV), which is equivalent to about 300 mg per serving. Rafferty et al., 2007, indicate the health claims:

- Foods that contain 10% of the DV for calcium, compared to a standard serving size of a similar food, may be said to be "calcium-enriched," "calcium-fortified," or have "more calcium."
- Foods containing 10% to 19% of the DV for calcium may say "Contains Calcium," "Provides Calcium," or is a "Good Source of Calcium."
- Foods containing 20% or more of the DV for calcium may be listed as "High in Calcium," "Rich in Calcium," or "Excellent Source of Calcium."

There are several calcium salts approved for fortifying foods, such as calcium carbonate, calcium chloride, calcium citrate, calcium hydroxide, and calcium lactate. The bioavailability of these salts, which varies between 20% and 40%, is influenced by factors such as solubility, source, and their interaction with food (Palacios et al., 2021; Bourassa et al., 2022). Such interactions can alter organoleptic aspects such as flavor, texture, smell, and color. Calcium fortification can increase acidity and bitterness, affecting the flavor of foods due to the associated calcium ions (Gómez-Álvarez et al., 2024). These changes depend on the type of salt used and the food matrix (Palacios et al., 2021). Although most calcium salts are soft, some, such as calcium citrate and calcium hydroxide, can have sour or slightly bitter flavors. High concentrations of calcium chloride and calcium lactate can be unpleasant (Trailokya et al., 2017). On the other hand, salts such as calcium carbonate, dicalcium phosphate, tribasic calcium phosphate, and calcium sulfate can impart a chalky taste and gritty mouthfeel, especially in large quantities (Nieto et al., 2021). Additionally, insoluble salts such as calcium carbonate can lighten the color of foods and cause sediment formation during storage, affecting the shelf life of liquid products. Due to these limitations, other sources of calcium are considered, such as eggshell, which is mainly composed of calcium carbonate (95%), magnesium carbonate (0.8%), and tricalcium phosphate (0.8%) (Brun et al., 2013; Bourassa et al., 2022).

Several studies have investigated the incorporation of eggshell powder into food products, such as calcium-fortified pork sausages, yogurt supplemented with nano eggshell powder (NPES), diet calcium biscuits, calcium-fortified roasted and ground coffee, calcium-enriched chocolate cake, pizza, pasta, cookies, fruit roll-ups, and dairy dessert (Daengprok et al., 2002; Al Mijan et al., 2014; Hassan et al., 2014; Aditya et al., 2021; Ray et al., 2017; Brun et al., 2013, Gómez et al., 2024).

Eggshells, due to their attractive properties, have been considered a valuable by-product to enrich food products, especially beneficial for people with osteoporosis or those in the growth stage, such as children. However, its use as a food additive faces technological challenges, especially regarding the impact on the structure and sensory properties of foods, which in the end are determining factors in the acceptance of the food and the shelf life of the product during storage (Kobus-Cisowska, et al., 2020). This highlights the difficulty in supplementing foods with eggshells, given their texture, granulometry, and safety. The development of new products requires a comprehensive approach that considers technological, dietary, and market aspects. It is essential to delve into studies that optimize both the technological and biological aspects, especially about the absorption and availability of calcium from the eggshell (Shkembi and Huppertz, 2021).

Under regulatory requirements for nutrient content claims, adequate levels of calcium in foods are established based on claims of calcium source (good) and high (excellent). Calcium source claims indicate the presence of the mineral at medium and high levels. A food that meets the definition of high or source of calcium is considered high in calcium (excellent source of calcium) or source of calcium (good source of calcium), respectively (Forouzesh et al., 2022).

References

Abdullah, N. H., Hasan, H., Razani, M. A. D., Hassan, M. H. I., Yaacob, M. S. S., Salim, N. A. A., & Yuan, N. S. (2022). *Phosphorus Removal from Aqueous Solution by Using Calcined Waste Chicken Eggshell: Kinetic and Isotherm Model.*

Aditya, S., Stephen, J., & Radhakrishnan, M. (2021). Utilization of eggshell waste in calcium-fortified foods and other industrial applications: A review. *Trends in Food Science & Technology,* 115, 422-432.

Aguirre, A., Gil-Quintana, E., Fenaux, M., Erdozain, S., & Sarria, I. (2017). Beneficial effects of oral supplementation with Ovoderm on human skin physiology: two pilot studies. *Journal of Dietary Supplements,* 14(6), 706-714.

Ahmed, T. A., Wu, L., Younes, M., & Hincke, M. (2021). Biotechnological applications of eggshell: recent advances. *Frontiers in Bioengineering and Biotechnology,* 9, 548.

Aina, S. T., Du Plessis, B. J., Mjimba, V., & Brink, H. G. (2022). *Eggshell valorization: membrane removal, calcium oxide synthesis, and biochemical compound recovery towards cleaner productions.*

Ajala, E. O., Eletta, O. A. A., Ajala, M. A., & Oyeniyi, S. K. (2018). Characterization and evaluation of chicken eggshell for use as a bio-resource. *Arid Zone Journal of Engineering, Technology and Environment,* 14(1), 26-40.

Al Mijan, M., Choi, K. H., & Kwak, H. S. (2014). Physicochemical, microbial, and sensory properties of nanopowdered eggshell-supplemented yogurt during storage. *Journal of Dairy Science,* 97(6), 3273-3280.

Al Omari, M. M. H., Rashid, I. S., Qinna, N. A., Jaber, A. M., & Badwan, A. A. (2016). Calcium carbonate. *Profiles of drug substances, excipients and related methodology,* 41, 31-132.

Amalraj, A., & Pius, A. (2015). Bioavailability of calcium and its absorption inhibitors in raw and cooked green leafy vegetables commonly consumed in India–An in vitro study. *Food chemistry,* 170, 430-436.

Andres, A., Quintana, E. G., & Manuel, L. N. (2018). Supplementation with Ovoderm® Reduces the Clinical Signs of Skin Aging. A Double-Blind, Placebo-Controlled Study. *Clin Res Dermatol Open Access,* 5(2), 1-8.

Apalangya, V., Rangari, V., Tiimob, B., Jeelani, S., & Samuel, T. (2014). Development of antimicrobial water filtration hybrid material from bio source calcium carbonate and silver nanoparticles. *Applied Surface Science,* 295, 108-114.

Arshad, R., Gulshad, L., Haq, I. U., Farooq, M. A., Al-Farga, A., Siddique, R., ... & Karrar, E. (2021). Nanotechnology: A novel tool to enhance the bioavailability of micronutrients. *Food Science & Nutrition,* 9(6), 3354-3361.

Baláž, P., Calka, A., Zorkovská, A., & Baláž, M. (2013). Processing of eggshell biomaterial by electrical discharge assisted mechanical milling (EDAMM) and high energy milling (HEM) techniques. *Materials and Manufacturing Processes,* 28(4), 343-347).

Baláž, M. (2014). Eggshell membrane biomaterial as a platform for applications in materials science. *Acta biomaterialia,* 10(9), 3827-3843.

Baláž, M. (2018). Ball milling of eggshell waste as a green and sustainable approach: a review. *Advances in colloid and interface science,* 256, 256-275.

Baláž, M., Boldyreva, E. V., Rybin, D., Pavlović, S., Rodríguez-Padrón, D., Mudrinić, T., & Luque, R. (2021). State-of-the-art of eggshell waste in materials science: recent advances in catalysis, pharmaceutical applications, and Mechanochemistry. *Frontiers in Bioengineering and Biotechnology,* 8, 612567.

Benkeblia, N. (Ed.). (2020). *Vitamins and Minerals Biofortification of Edible Plants*. John Wiley & Sons.

Bourassa, M. W., Abrams, S. A., Belizán, J. M., Boy, E., Cormick, G., Quijano, C. D., ... & Weaver, C. M. *Interventions to improve calcium intake through foods in populations with low intake,* https://doi.org/10.1111/nyas.14743, Ann. New York Acad. Sci., 1511(1), (2022), pp. 40-58.

Brun, L. R., Lupo, M., Delorenzi, D. A., Di Loreto, V. E., & Rigalli, A. (2013). Chicken eggshell as suitable calcium source at home. *International journal of food sciences and nutrition,* 64(6), 740-743.

Cao, Y. (2022). Nutrient molecule corona: an update for nanomaterial-food component interactions. *Toxicology,* 153253.

Cha, S. W., Yoon, J. D., & Uto, N. (2004). A study on the effect of porous CaCO3 on micro-cellular plastics as an additive for nucleation. *Cellular polymers,* 23(4), 229-241.

Chen, N., Wan, C., Zhang, Y., & Zhang, Y. (2004). Effect of nano-CaCO3 on mechanical properties of PVC and PVC/Blendex blend. *Polymer Testing,* 23(2), 169-174.

Chen, J., Jiang, M., Su, M., Han, J., Li, S., & Wu, Q. (2019). Mineralization of calcium carbonate induced by egg substrate and an electric field. *Chemical Engineering & Technology, 42*(7), 1525-1532.

Chervonnyi, A. D., Chervonnaya, N. A., & Chukanov, N. V. (2003). Effect of CaCO3 polymorphs on the strength of a calcium aluminosilicate composite. *Inorganic materials,* 39(4), 386-391.

Cölfen, H. (2003). Precipitation of carbonates: recent progress in controlled production of complex shapes. *Current opinion in colloid & interface science,* 8(1), 23-31.

Cordeiro, C. y T Hincke, M. (2011). Recent patents on eggshell: shell and membrane applications. *Recent patents on food, nutrition & agriculture,* 3(1), 1-8.

D'Haese, M., Langouche, F., & Van Puyvelde, P. (2013). On the effect of particle size, shape, concentration, and aggregation on the flow-induced crystallization of polymers. *Macromolecules,* 46(9), 3425-3434.

Dhami, N. K., Reddy, M. S., & Mukherjee, A. (2013). Biomineralization of calcium carbonates and their engineered applications: a review. *Frontiers in microbiology,* 4, 314.

da Silveira Pinto, L. S., & de Souza, M. V. (2019). Eggshell, a promising waste in organic reactions. *Letters in Organic Chemistry,* 16(11), 851-859.

de Paula, H. A., Becker, J. G., & Davis, A. P. (2008). Characterization of the uptake of divalent metal ions by a hatchery residual. *Environmental engineering science,* 25(5), 737-746.

Daengprok, W., Garnjanagoonchorn, W., & Mine, Y. (2002). Fermented pork sausage fortified with commercial or hen eggshell calcium lactate. *Meat science,* 62(2), 199-204.

Das, A. K., Islam, M., Ashaduzzaman, M., & Nazhad, M. M. (2020). Nanocellulose: its applications, consequences and challenges in papermaking. *Journal of Packaging Technology and Research,* 4(3), 253-260.

Dash, K. K., Deka, P., Bangar, S. P., Chaudhary, V., Trif, M., & Rusu, A. (2022). Applications of inorganic nanoparticles in food packaging: A Comprehensive Review. *Polymers,* 14(3), 521.

Duncan, T. V. (2011). Applications of nanotechnology in food packaging and food safety: barrier materials, antimicrobials and sensors. *Journal of colloid and interface science,* 363(1), 1-24.

Dwyer, J. T., Wiemer, K. L., Dary, O., Keen, C. L., King, J. C., Miller, K. B., ... & Bailey, R. L. (2015). Fortification and health: challenges and opportunities. *Advances in nutrition*, *6*(1), 124-131.

EGGNOVO (2016). Las soluciones Eggnovo superan sus primeros test precomercialización. *Navarra Capital.* (Recuperado el 20 de febrero de 2024). Disponible en: https://navarracapital.es/las-soluciones-eggnovo-superan-susprimeros-test-pre-comercializacion/

Elemike, E. E., Ekennia, A. C., Onwudiwe, D. C., & Ezeani, R. O. (2022). Agro-waste materials: Sustainable substrates in nanotechnology. In *Agri-Waste and Microbes for Production of Sustainable Nanomaterials* (pp. 187-214). Elsevier.

Erfanian, A., Mirhosseini, H., Abd Manap, M. Y., Rasti, B., & Bejo, M. H. (2014). Influence of nano-size reduction on absorption and bioavailability of calcium from fortified milk powder in rats. *Food research international,* 66, 1-11.

Erfanian, A., Rasti, B., & Manap, Y. (2017). Comparing the calcium bioavailability from two types of nano-sized enriched milk using in-vivo assay. *Food Chemistry,* 214, 606-613.

Ersoy, O., Güler, D., & Rençberoğlu, M. (2021). Effects of Grinding Aids Used in Grinding Calcium Carbonate (CaCO3) Filler on the Properties of Water-Based Interior Paints. *Coatings,* 12(1), 44.

Fadia, P., Tyagi, S., Bhagat, S., Nair, A., Panchal, P., Dave, H., ... & Singh, S. (2021). Calcium carbonate nano-and microparticles: synthesis methods and biological applications. *3 Biotech,* 11(11), 1-30.

Faridi, H., & Arabhosseini, A. (2018). Application of eggshell wastes as valuable and utilizable products: A review. *Research in Agricultural Engineering,* 64(2), 104-114.

Fernandes, M., Almeida Paz, F. A., Mano, J. F., & de Zea Bermudez, V. (2014). Investigation of calcium carbonate precipitated in the presence of alkanols. *Crystal Research and Technology,* 49(6), 418-430.

Forouzesh, A., Forouzesh, F., Samadi Foroushani, S., Forouzesh, A., & Zand, E. (2022). A new method for calculating calcium content and determining appropriate calcium levels in foods. *Food Analytical Methods*, 1-10.

Fu, L. H., Dong, Y. Y., Ma, M. G., Yue, W., Sun, S. L., & Sun, R. C. (2013). Why to synthesize vaterite polymorph of calcium carbonate on the cellulose matrix via sonochemistry process. *Ultrasonics sonochemistry,* 20(5), 1188-1193.

Fu, J., Leo, C. P., & Show, P. L. (2022). Recent advances in the synthesis and applications of pH responsive $CaCO_3$. *Biochemical Engineering Journal,* 108446.

Fu, Q., Zhang, Z., Zhao, X., Xu, W., & Niu, D. (2022). Effect of nano calcium carbonate on hydration characteristics and microstructure of cement-based materials: A review. *Journal of Building Engineering,* 50, 104220.

Gandolfi, M. G., Siboni, F., Botero, T., Bossù, M., Riccitiello, F., & Prati, C. (2015). Calcium silicate and calcium hydroxide materials for pulp capping: biointeractivity, porosity, solubility and bioactivity of current formulations. *Journal of applied biomaterials & functional materials,* 13(1), 43-60.

Gautron, J., Hincke, M. T., & Nys, Y. (1997). Precursor matrix proteins in the uterine fluid change with stages of eggshell formation in hens. *Connective tissue research,* 36(3), 195-210.

Gil-Quintana, E., & Molero, A. (2020). Ovopet a new and effective treatment to decrease inflammation, pain and lameness in competing trotters. *Journal of Veterinary Medicine and Animal Health,* 12(1), 1-6.

Gómez-Alvarez, Luz M., Segura-Sánchez, Freimar, & Zapata, José E. (2022). Combination of high shear and ultrasound to obtain calcium carbonate nanoparticles from eggshells. *Información tecnológica,* 33(1), 91-106. https://dx.doi.org/10.4067/S0718-07642022000100091.

Gómez-Alvarez, L. M., & Montoya, J. E. Z. (2024). Effect of fortification with $CaCO_3$ nanoparticles obtained from eggshell on the physical and sensory characteristics of three food matrices. *Heliyon,* 10(2).

Gómez-Llorente, H., Hervás, P., Pérez-Esteve, É., Barat, J. M., & Fernández-Segovia, I. (2022). Nanotechnology in the agri-food sector: Consumer perceptions. *NanoImpact,* 26, 100399.

González, V., & Patricia, M. (2020). Potential uses of hen eggshells gallina (*Gallus gallus domesticus*): a systemic review. *Revista colombiana de ciencia animal recia,* 12(2), 106-116.

Guijarro-Aldaco, A., Hernández-Montoya, V., Bonilla-Petriciolet, A., Montes-Morán, M. A., & Mendoza-Castillo, D. I. (2011). Improving the adsorption of heavy metals from water using commercial carbons modified with eggshell wastes. *Industrial & engineering chemistry research,* 50(15), 9354-9362.

Guru, P. S., & Dash, S. (2014). Sorption on eggshell waste—a review on ultrastructure, biomineralization and other applications. *Advances in colloid and interface science,* 209, 49-67.

Hamideh, F., & Akbar, A. (2018). Application of eggshell wastes as valuable and utilizable products: A review. *Research in Agricultural Engineering,* 64(2), 104-114.

Hassan, T. A., Rangari, V. K., Rana, R. K., & Jeelani, S. (2013). Sonochemical effect on size reduction of CaCO3 nanoparticles derived from waste eggshells. *Ultrasonics sonochemistry,* 20(5), 1308-1315.

Hassan, T. A., Rangari, V. K., & Jeelani, S. (2014). Value-added biopolymer nanocomposites from waste eggshell-based $CaCO_3$ nanoparticles as fillers. *ACS Sustainable Chemistry & Engineering,* 2(4), 706-717.

Hassan, N. M. (2015). Chicken eggshell powder as dietary calcium source in biscuits. *World J. Dairy Food Sci,* 10(2), 199-206.

He, X., & Hwang, H. M. (2016). Nanotechnology in food science: Functionality, applicability, and safety assessment. *Journal of food and drug analysis,* 24(4), 671-681.

Heaney, R. P. (2000). Calcium, dairy products and osteoporosis. *Journal of the American college of nutrition,* 19(sup2), 83S-99S.

Hincke, M. T., Gautron, J., Panheleux, M., Garcia-Ruiz, J., McKee, M. D., & Nys, Y. (2000). Identification and localization of lysozyme as a component of eggshell membranes and eggshell matrix. *Matrix Biology,* 19(5), 443-453.

Hincke, M. T., Nys, Y., Gautron, J., Mann, K., Rodriguez-Navarro, A. B., & McKee, M. D. (2012). The eggshell: structure, composition and mineralization. *Frontiers in Bioscience-Landmark,* 17(4), 1266-1280.

Huang, S., Chen JC, Hsu CW, Chang WH. (2009). Effects of nano calcium carbonate and nano calcium citrate on toxicity in ICR mice and on bone mineral density in an ovariectomized mice model. *Nanotechnology,* Volume 20, Number 37.

Huang, X., Dong, K., Liu, L., Luo, X., Yang, R., Song, H., ... & Huang, Q. (2020). Physicochemical and structural characteristics of nano eggshell calcium prepared by wet ball milling. *LWT,* 131, 109721.

Ibrahim, A. R., Vuningoma, J. B., Huang, Y., Wang, H., & Li, J. (2014). Rapid carbonation for calcite from a solid-liquid-gas system with an imidazolium-based ionic liquid. *International Journal of Molecular Sciences,* 15(7), 11350-11363.

Jagtiani, E. (2022). Advancements in nanotechnology for food science and industry. *Food Frontiers,* 3(1), 56-82.

Jeong, M. S., Cho, H. S., Park, S. J., Song, K. S., Ahn, K. S., Cho, M. H., & Kim, J. S. (2013). Physico-chemical characterization-based safety evaluation of nanocalcium. *Food and chemical toxicology,* 62, 308-317.

Jimoh, O. A., Ariffin, K. S., Hussin, H. B., & Temitope, A. E. (2018). Synthesis of precipitated calcium carbonate: a review. *Carbonates and Evaporites,* 33(2), 331-346.

Kamali, M., Appels, L., Kwon, E. E., Aminabhavi, T. M., & Dewil, R. (2021). Biochar in water and wastewater treatment-a sustainability assessment. *Chemical Engineering Journal,* 420, 129946.

Kasmuri, N., & Zait, M. S. A. (2018). Enhancement of bio-plastic using eggshells and chitosan on potato starch based. *Int. J. Eng. Technol,* 7, 110-115.

Karakaş, F., Hassas, B. V., & Celik, M. S. (2015). Effect of precipitated calcium carbonate additions on waterborne paints at different pigment volume concentrations. *Progress in Organic Coatings,* 83, 64-70.

Kavitha, V., Geetha, V., & Jacqueline, P. J. (2019). Production of biodiesel from dairy waste scum using eggshell waste. *Process Safety and Environmental Protection,* 125, 279-287.

Kim, M. K., Lee, J. A., Jo, M. R., Kim, M. K., Kim, H. M., Oh, J. M., ... & Choi, S. J. (2015). Cytotoxicity, uptake behaviors, and oral absorption of food grade calcium carbonate nanomaterials. *Nanomaterials,* 5(4), 1938-1954.

Kobus-Cisowska, J., Szymanowska-Powałowska, D., Szymandera-Buszka, K., Rezler, R., Jarzębski, M., Szczepaniak, O., ... & Kobus-Moryson, M. (2020). Effect of fortification with calcium from eggshells on bioavailability, quality, and rheological characteristics of traditional Polish bread spread. *Journal of dairy science,* 103(8), 6918-6929.

Kopac, T. (2021). Protein corona, understanding the nanoparticle–protein interactions and future perspectives: A critical review. *International Journal of Biological Macromolecules,* 169, 290-301.

Krammer, G., Gasparin, G., Staudinger, G., & Niederkofler, R. (2002). Formation of Calcium Carbonate Sub-Micron Particles in a High Shear Stress Three-Phase Reactor. *Particle & Particle Systems Characterization: Measurement and Description of Particle Properties and Behavior in Powders and Other Disperse Systems,* 19(5), 348-353.

Kumar, P., Mahajan, P., Kaur, R., & Gautam, S. (2020). Nanotechnology and its challenges in the food sector: A review. *Materials Today Chemistry,* 17, 100332.

Leach Jr, R. M. (1982). Biochemistry of the organic matrix of the eggshell. *Poultry Science,* 61(10), 2040-2047.

Le Roy, N., Stapane, L., Gautron, J., & Hincke, M. T. (2021). Evolution of the Avian Eggshell Biomineralization Protein Toolkit–New Insights From Multi-Omics. *Frontiers in genetics,* 12, 672433.

Lee, S. W., Kim, S. G., Balázsi, C., Chae, W. S., & Lee, H. O. (2012). Comparative study of hydroxyapatite from eggshells and synthetic hydroxyapatite for bone regeneration. *Oral surgery, oral medicine, oral pathology and oral radiology,* 113(3), 348-355.

Lewko, L., Krawczyk, J., & Calik, J. (2021). Effect of genotype and some shell quality traits on lysozyme content and activity in the albumen of eggs from hens under the biodiversity conservation program. *Poultry Science,* 100(3), 100863.

Lewicka, E., Szlugaj, J., Burkowicz, A., & Galos, K. (2020). Sources and markets of limestone flour in Poland. *Resources,* 9(10), 118.

Liao, D., Zheng, W., Li, X., Yang, Q., Yue, X., Guo, L., & Zeng, G. (2010). Removal of lead (II) from aqueous solutions using carbonate hydroxyapatite extracted from eggshell waste. *Journal of hazardous materials,* 177(1-3), 126-130.

Lieder, M., & Rashid, A. (2016). Towards circular economy implementation: a comprehensive review in context of manufacturing industry. *Journal of cleaner production,* 115, 36-51.

Mattar, G., Haddarah, A., Haddad, J., Pujola, M., & Sepulcre, F. (2022). New approaches, bioavailability and the use of chelates as a promising method for food fortification. *Food Chemistry,* 373, 131394.

McClements, D. J., DeLoid, G., Pyrgiotakis, G., Shatkin, J. A., Xiao, H., & Demokritou, P. (2016). The role of the food matrix and gastrointestinal tract in the assessment of biological properties of ingested engineered nanomaterials (iENMs): State of the science and knowledge gaps. *NanoImpact,* 3, 47-57.

McClements, D. J. (2020). Recent advances in the production and application of nano-enabled bioactive food ingredients. *Current opinion in food science,* 33, 85-90.

Market, P. (2020). *Forecast* (2020–2025). Report Linker.

Martin-Luengo, M. A., Yates, M., Ramos, M., Salgado, J. L., Aranda, R. M., Plou, F., ... & Ruiz-Hitzky, E. (2011, November). Renewable raw materials for advanced applications. In 2011 World Congress on Sustainable Technologies (WCST) (pp. 19-22). *IEEE.*

Mignardi, S., Archilletti, L., Medeghini, L., & De Vito, C. (2020). Valorization of eggshell biowaste for sustainable environmental remediation. *Scientific Reports,* 10(1), 1-10.

Mony, B., Ebenezar, A. R., Ghani, M. F., Narayanan, A., Anand, S., & Mohan, A. G. (2015). Effect of chicken eggshell powder solution on early enamel carious lesions: an invitro preliminary study. *Journal of clinical and diagnostic research: JCDR,* 9(3), ZC30.

Moradi, M., Razavi, R., Omer, A. K., Farhangfar, A., & McClements, D. J. (2022). Interactions between nanoparticle-based food additives and other food ingredients: A review of current knowledge. *Trends in Food Science & Technology.*

Moriconi, L., Nascimento, T., de Souza, B. G. B., & Loureiro, J. B. R. (2022). Top-down model of calcium carbonate scale formation in turbulent pipe flows. *Thermal Science and Engineering Progress,* 28, 101141.

Muhammad, A. F., Abidin, M. S. Z., Hassan, M. H., Mustafa, Z., & Anjang, A. (2022). *Effect of eggshell fillers on the tensile and flexural properties of glass fiber reinforced polymer composites.* Materials Today: Proceedings.

Murakami, F. S., Rodrigues, P. O., Campos, C. M. T. D., & Silva, M. A. S. (2007). Physicochemical study of $CaCO_3$ from eggshells. *Food Science and Technology,* 27, 658-662.

Nawaz, A., Li, E., Irshad, S., Xiong, Z., Xiong, H., Shahbaz, H. M., & Siddique, F. (2020). Valorization of fisheries by-products: Challenges and technical concerns to food industry. *Trends in Food Science & Technology,* 99, 34-43.

Nieto, J. A., Soriano-Romaní, L., Tomás-Cobos, L., Sharma, L., & Budde, T. (2021). Improved in vitro bioavailability of a newly developed functionalized calcium carbonate salt as a food ingredient and its comparison with available commercial calcium salts. *Food Chemistry,* 348, 128740.

Nile, S. H., Baskar, V., Selvaraj, D., Nile, A., Xiao, J., & Kai, G. (2020). Nanotechnologies in food science: applications, recent trends, and future perspectives. *Nano-micro letters,* 12(1), 1-34.

Nyamaizi, S., Messiga, A. J., Cornelis, J. T., & Smukler, S. M. (2022). Effects of increasing soil pH to near-neutral using lime on phosphorus saturation index and water extractable phosphorus. *Canadian Journal of Soil Science,* (ja).

Nys, Y., Zawadzki, J., Gautron, J., & Mills, A. D. (1991). Whitening of brown-shelled eggs: mineral composition of uterine fluid and rate of protoporphyrin deposition. *Poultry Science,* 70(5), 1236-1245.

Nys, Y. (2001). La coquille d'œuf: un biomatériau composite. *Pour la science,* (289), 48-54.

Nys, Y., Gautron, J., Garcia-Ruiz, J. M., & Hincke, M. T. (2004). Avian eggshell mineralization: biochemical and functional characterization of matrix proteins. *Comptes Rendus Palevol,* 3(6-7), 549-562.

Nuhu, A. A., Basheer, C., Shaikh, A. A., & Al-Arfaj, A. R. (2012). Determination of polycyclic aromatic hydrocarbons in water using nanoporous material prepared from waste avian eggshell. *Journal of Nanomaterials,* 2012.

Ok, Y. S., Lee, S. S., Jeon, W. T., Oh, S. E., Usman, A. R., & Moon, D. H. (2011). Application of eggshell waste for the immobilization of cadmium and lead in a contaminated soil. Environmental geochemistry and health, 33(1), 31-39.

Oliveira, D. A., Benelli, P., & Amante, E. R. (2013). A literature review on adding value to solid residues: Egg shells. *Journal of cleaner production,* 46, 42-47.

Oriols, N., Salvadó, N., Pradell, T., & Butí, S. (2020). Amorphous calcium carbonate (ACC) in fresco mural paintings. *Microchemical Journal,* 154, 104567.

Palacios, C., Cormick, G., Hofmeyr, G. J., Garcia-Casal, M. N., Peña-Rosas, J. P., & Betrán, A. P. (2021). Calcium-fortified foods in public health programs: considerations for implementation. *Annals of the New York Academy of Sciences*, *1485*(1), 3-21.

Parakhonskiy, B., Haase, A., Piccoli, F., Caola, I., Tessarolo, F., Bukreeva, T., & Antolini, R. (2011). *Porous vaterite particles as drug delivery system: synthesis, encapsulation, and controlled release*.

Park, K., Sadeghi, K., Thanakkasaranee, S., Park, Y. I., Park, J., Nam, K. H., ... & Seo, J. (2021). Effects of calcination temperature on morphological and crystallographic properties of oyster shell as biocidal agent. *International Journal of Applied Ceramic Technology,* 18(2), 302-311.

Platon, N., Georgescu, A. M., Aruş, V. A., Sion, I., Silion, M., Ursu, A. V., & Nistor, I. D. (2022). *Valorization Of Eggshells Waste For Bread Production.*

Polat, S., & Sayan, P. (2020). Ultrasonic-assisted eggshell extract-mediated polymorphic transformation of calcium carbonate. *Ultrasonics Sonochemistry,* 66, 105093.

Pokorski, P., & Hoffmann, M. (2022). Valorization of bio-waste eggshell as a viable source of dietary calcium for confectionery products. *Journal of the Science of Food and Agriculture,* 102(8), 3193-3203.

Pushparaj, K., Liu, W. C., Meyyazhagan, A., Orlacchio, A., Pappusamy, M., Vadivalagan, C., ... & Balasubramanian, B. (2022). Nano-from nature to nurture: A comprehensive review on facets, trends, perspectives and sustainability of nanotechnology in the food sector. *Energy,* 240, 122732.

Quddoos, M. Y., Mahmood, S., Yaqoob, M., Murtaza, M. A., Zahra, S. M., ud Din, G. M., ... & Mustafa, S. (2022). The effects of natural and synthetic calcium utilization on quality parameters of cookies. *Applied Food Research,* 2(1), 100093.

Quina, M. J., Soares, M. A., & Quinta-Ferreira, R. (2017). Applications of industrial eggshell as a valuable anthropogenic resource. *Resources, Conservation and Recycling,* 123, 176-186.

Rafferty, K., Walters, G., & Heaney, R. P. (2007). Calcium fortificants: overview and strategies for improving calcium nutriture of the US population. *Journal of food science,* 72(9), R152-R158.

Ray, S., Barman, A. K., Roy, P. K., & Singh, B. K. (2017). Chicken eggshell powder as dietary calcium source in chocolate cakes. *The Pharma Innovation,* 6(9, Part A), 1.

Rivera, A. M. P., Figueroa, A. G., Toro, C. R., & Bolivar, G. (2022). Adding value to processes, products, and by-products from the poultry industry through enzymatic technologies. In *Value-Addition in Food Products and Processing Through Enzyme Technology* (pp. 235-251). Academic Press.

Rohim, R., Ahmad, R., Ibrahim, N., Hamidin, N., & Abidin, C. Z. A. (2014). Characterization of calcium oxide catalyst from eggshell waste. *Adv Environ Biol,* 8(22), 35-38.

Sabir, A., Altaf, F., Batool, R., Shafiq, M., Khan, R. U., & Jacob, K. I. (2021). Agricultural waste absorbents for heavy metal removal. In *Green Adsorbents to Remove Metals, Dyes and Boron from Polluted Water* (pp. 195-228). Springer, Cham.

Sah, N., & Mishra, B. (2018). Regulation of egg formation in the oviduct of laying hen. *World's Poultry Science Journal,* 74(3), 509-522.

Sanguansri, P., & Augustin, M. A. (2006). Nanoscale materials development–a food industry perspective. *Trends in Food Science & Technology,* 17(10), 547-556.

Santillán-Urquiza, E., Méndez-Rojas, M. E., Vélez-Ruiz, J. F., Fortification of yogurt with nano and micro sized calcium, iron and zinc, effect on the physicochemical and rheological properties, *LWT - Food Science and Technology* (2017), doi: 10.1016/j.lwt.2017.03.025.

Santos, R. M., Ceulemans, P., & Van Gerven, T. (2012). Synthesis of pure aragonite by sonochemical mineral carbonation. *Chemical Engineering Research and Design,* 90(6), 715-725.

Setyawati, M. I., Zhao, Z., & Ng, K. W. (2020). Transformation of nanomaterials and its implications in gut nanotoxicology. *Small,* 16(36), 2001246.

Schaafsma, A., Pakan, I., Hofstede, G. J. H., Muskiet, F. A., Van Der Veer, E., & De Vries, P. J. F. (2000). Mineral, amino acid, and hormonal composition of chicken eggshell powder and the evaluation of its use in human nutrition. *Poultry science,* 79(12), 1833-1838.

Shang, B., Wang, S., Lu, L., Ma, H., Liu, A., Zupanic, A., ... & Yu, Y. (2022). Poultry eggshell-derived antimicrobial materials: Current status and future perspectives. *Journal of Environmental Management,* 314, 115096.

Shen, S., Wu, Y., Liu, Y., & Wu, D. (2017). High drug-loading nanomedicines: progress, current status, and prospects. *International Journal of nanomedicine,* 12, 4085.

Shimpi, N., Mali, A., Hansora, D. P., & Mishra, S. (2015). Synthesis and surface modification of calcium carbonate nanoparticles using ultrasound cavitation technique. *Nanoscience and Nanoengineering,* 3(1), 8-12.

Shkembi, B., & Huppertz, T. (2021). Calcium absorption from food products: food matrix effects. *Nutrients,* 14(1), 180.

Schröder, R., Pohlit, H., Schüler, T., Panthöfer, M., Unger, R. E., Frey, H., & Tremel, W. (2015). Transformation of vaterite nanoparticles to hydroxycarbonate apatite in a hydrogel scaffold: Relevance to bone formation. *Journal of Materials Chemistry B,* 3(35), 7079-7089.

Sonawane, S. H., Kapadnis, C. V., Meshram, S., Gumfekar, S. P., & Khanna, P. K. (2010). Sonochemical formation of CaCO3 nanoparticles with controlled particle size distribution. *International Journal of Green Nanotechnology: Physics and Chemistry,* 2(2), P69-P79.

Stapane, L., Le Roy, N., Ezagal, J., Rodriguez-Navarro, A. B., Labas, V., Combes-Soia, L., ... & Gautron, J. (2020). Avian eggshell formation reveals a new paradigm for vertebrate mineralization via vesicular amorphous calcium carbonate. *Journal of Biological Chemistry,* 295(47), 15853-15869.

Stoica-Guzun, A., Stroescu, M., Jinga, S., Jipa, I., Dobre, T., & Dobre, L. (2012). Ultrasound influence upon calcium carbonate precipitation on bacterial cellulose membranes. *Ultrasonics Sonochemistry,* 19(4), 909-915.

Summa, D., Lanzoni, M., Castaldelli, G., Fano, E. A., & Tamburini, E. (2022). Trends and Opportunities of Bivalve Shells' Waste Valorization in a Prospect of Circular Blue Bioeconomy. *Resources,* 11(5), 48.

Trailokya, A., Srivastava, A., Bhole, M., & Zalte, N. (2017). Calcium and calcium salts. *Journal of the Association of Physicians of India*, *65*(1), 1-2.

Tutus, A., Killi, U., & Cicekler, M. (2020). Evaluation of eggshell wastes in office paper production. *Biomass Conversion and Biorefinery,* 1-10.

Vandeginste, V. (2021). Food waste eggshell valorization through development of new composites: A review. *Sustainable Materials and Technologies,* 29, e00317.

Vilarinho, I. S., Filippi, E., & Seabra, M. P. (2022). Bio-CaCO3 from Eggshell Waste as Raw Material for Eco-Ceramic Products. *Materials Proceedings,* 8(1), 58.

Waheed, M., Butt, M. S., Shehzad, A., Adzahan, N. M., Shabbir, M. A., Suleria, H. A. R., & Aadil, R. M. (2019). Eggshell calcium: A cheap alternative to expensive supplements. *Trends in Food Science & Technology,* 91, 219-230.

Waheed, M., Yousaf, M., Shehzad, A., Inam-Ur-Raheem, M., Khan, M. K. I., Khan, M. R., ... & Aadil, R. M. (2020). Channelling eggshell waste to valuable and utilizable products: a comprehensive review. *Trends in Food Science & Technology,* 106, 78-90.

Wang, L., Manson, J. E., & Sesso, H. D. (2012). Calcium intake and risk of cardiovascular disease. *American Journal of Cardiovascular Drugs,* 12(2), 105-116.

Wawrzyniak, N., & Suliburska, J. (2021). Nutritional and health factors affecting the bioavailability of calcium: a narrative review. *Nutrition Reviews*, *79*(12), 1307-1320.

Wei, Z., Xu, C., & Li, B. (2009). Application of waste eggshell as low-cost solid catalyst for biodiesel production. *Bioresource technology,* 100(11), 2883-2885.

Wellman-Labadie, O., Picman, J., & Hincke, M. T. (2008). Antimicrobial activity of cuticle and outer eggshell protein extracts from three species of domestic birds. *British poultry science,* 49(2), 133-143.

Wilson, P. B. (2017). Recent advances in avian egg science: A review. *Poultry science,* 96(10), 3747-3754.

Wu, X., Stroll, S. I., Lantigua, D., Suvarnapathaki, S., & Camci-Unal, G. (2019). Eggshell particle-reinforced hydrogels for bone tissue engineering: An orthogonal approach. *Biomaterials science,* 7(7), 2675-2685.

Xiao, N., Huang, X., He, W., Yao, Y., Wu, N., Xu, M., ... & Tu, Y. (2021). A review on recent advances of egg byproducts: Preparation, functional properties, biological activities and food applications. *Food Research International,* 147, 110563.

Xiang, L., Xiang, Y., Wang, Z. G., & Jin, Y. (2002). Influence of chemical additives on the formation of super-fine calcium carbonate. *Powder Technology,* 126(2), 129-133.

Yang, X., & Xu, G. (2011). The influence of xanthan on the crystallization of calcium carbonate. *Journal of Crystal Growth,* 314(1), 231-238.

Yang, A., Huang, Z., Zhu, Y., Han, Y., & Tong, Z. (2021). Preparation of nano-sized calcium carbonate in solution mixing process. *Journal of Crystal Growth,* 571, 126247.

Zárybnická, L., Ševčík, R., Pokorný, J., Machová, D., Stránská, E., & Šál, J. (2022). CaCO3 polymorphs used as additives in filament production for 3D printing. *Polymers,* 14(1), 199.

Zhang, X., Fan, Z., Lu, Q., Huang, Y., Kaplan, D. L., & Zhu, H. (2013). Hierarchical biomineralization of calcium carbonate regulated by silk microspheres. *Acta biomaterialia,* 9(6), 6974-6980.

Zhang, J., Li, G., Xu, D., & Cao, Y. (2021). Stability, Microstructure, and Rheological Properties of CaCO3 S/O/W Calcium-Lipid Emulsions. *Foods,* 10(9), 2216.

Zhou, J., Wang, S., Nie, F., Feng, L., Zhu, G., & Jiang, L. (2011). Elaborate architecture of the hierarchical hen's eggshell. *Nano Research*, 4(2), 171-179.

Zhou, H., Wang, X., Wang, T., Jian, J., Zhou, Z., Zeng, J., ... & Liu, G. (2019). Polyurethane synthetic papers based on different inorganic fillers with water and fire resistance. *Macromolecular Materials and Engineering,* 304(3), 1800473.

Chapter 4

Red Tilapia (*Oreochromis* spp.) Viscera Silage as a Source of Animal Protein, a Profitable Alternative: A Technical-Economic Study

José E. Zapata*, PhD
Jairo A. Camaño, PhD Candidate
and Yhoan S. Gaviria, PhD Candidate
Department of Food and Pharmaceutical Sciences, Food Nutrition and Technology Group, Universidad de Antioquia, Medellín, Colombia

Abstract

An analysis to determine the economic feasibility of using Red Tilapia (*Oreochromis* spp.) viscera for extracting oil and producing dry chemical silage (DCS). A conventional Red Tilapia (*Oreochromis* spp.) production farm (FFC) was used as a base, and a farm that uses viscera to obtain oil and DCS (FFS) was used as a pilot. The economic parameters of investment, production, balance points (BP), and financial indicators such as Net Present Value, Internal Return Rate, Cost-Benefit Ratio (CBR), and Investment Payback Period (IPP) were determined. Both systems showed economic feasibility in the study. However, the system that uses the viscera showed better economic indicators thanks to oils and DCS sales, therefore showing that such activity has practical implementation potential.

Keywords: chemical silage, balance point, net present value, internal return rate, cost-benefit ratio

---

* Corresponding Author's E-mail: edgar.zapata@udea.edu.co.

In: Agro-Industrial Wastes
Editor: Peter Clements
ISBN: 979-8-89113-688-5

Introduction

Fish by-products silage is one of the solution alternatives used to mitigate the growing problem of waste produced in this industry (Van't Land and Raes, 2019). Thus, silage has been used to process trout waste (Goosen et al., 2014), red tilapia (Camaño et al., 2021), and other species (Davies et al., 2020). Not only do these products offer environmental benefits due to waste reduction, but also show to be a protein substrate with great benefits for the nutrition of different farm animals such as trout (Güllü et al., 2014), broiler chickens (Gómez et al., 2014; Gaviria et al., 2021) and laying hens (Gaviria et al., 2020), among other animal species (Goosen et al., 2014). Recently produced silage is obtained in liquid stare, which may cause some conservation problems, mostly due to fungi formation because of its high water content (Madage et al., 2015; Camaño et al., 2020). Therefore, to improve conservation, a drying stage must be included (Camaño et al., 2021). An option that has been implemented with positive results is sun drying, as it is environmentally friendly and does not involve the high expenses of other drying processes. (Camaño et al., 2021). Obtaining dried silage using solar drying comes as a low-cost, environmentally friendly alternative that is easy to implement in rural areas, since it does not require any high-end technology (Suarez et al., 2018).

Nevertheless, there are few studies that assess the technical and economic feasibility of implementing this type of process on a productive scale. This study carries out a technical-economic study of dry silage production from red tilapia (*Oreochromis* spp) viscera, aiming to provide fish farmers with information that allows them to tackle this kind of initiatives at a production level.

Methodology

This study was based on data from small fish farmers in Colombia, who have small productions at an individual level but generate a high volume of waste altogether. Some of the most important waste is viscera, whose inadequate disposal may cause important environmental impacts, as well as significant costs (Villamil et al., 2017).

This methodology has several subsections. First, it provides details on the systems and processes included in this study, followed by cost estimation, balance points, net income and revenue, and finally the cost-benefit ratio, which includes the financial indicators NPV, IRR, CBR, and IPP.

Study Systems

Two production systems were proposed, one of which was based on the industrial use of waste resulting from the evisceration of a red tilapia (*Oreochromis* spp) farm (FFS) from pilot studies carried out previously in the same group that presents this study, whereas the other system consisted of a conventional fish production system, used as the control system. Each system's description is shown in Figure 1.

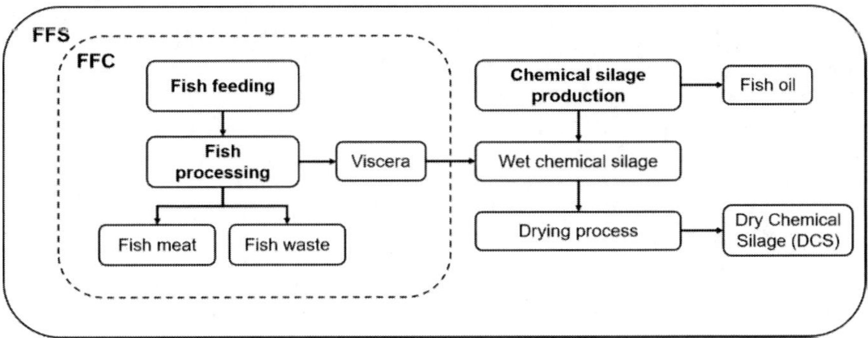

Figure 1. Diagram of the both production systems.

Fish Farming

As a calculation base for fish production, 15,000 red tilapia (*Oreochromis* spp.) fingerlings were used and raised for 7 months. Considering a mortality rate of 13% (Gonzales *and* Quevedo, 2018; FAO, 2009), 13.050 adult fish were obtained. Pre-fattening and fattening stages were developed based on technical and productive parameters taken from the FAO (FAO, 2009).

The system consisted of 3 tanks with geomembranes for the pre-fattening stage for 3 months (50 fish/m^3, diameter 4.2 m), and 4 tanks for the fattening stage for 4 months (50 fish/m^3, diameter 6.8m). 1.864 fish/month were obtained (0.40 kg/fish) for a total weight of 745.6 kg, which gave 119,314 kg of viscera/month (16%) (Villamil et al., 2017; Suarez et al.,

2018), whose disposal depended on the type of farm that calculated it according to the case.

Dry Silage Production

Viscera produced by the evisceration of fish from the farming are taken to a degreasing process following the methodology described by Arias et al., 2017. Then the grease is stored for further use, and the protein phase is taken to the silage process, carried out according to the methodology described by Suarez et al., 2018. Wet silage is dehydrated in solar dryers as described by Camaño et al., 2020 and Gaviria et al., 2021, producing a total of 109 kg of DCS and 247.52 kg of oil per cycle from 100% of the viscera used. In the case of the FFS system, the dry CS (DCS) and the fish oil are used for sale.

Costs

Costs were divided into 3 categories: (1) investment costs, (2) variable costs, and (3) fixed costs. Cost estimation for each type of farm was carried out over 7-month productive cycles. Investment costs were calculated in both systems (1) as the sum of the cost of materials, supplies and equipment needed to start the production cycle (Weber et al., 2020), keeping into consideration the information of suppliers and quotes for budget, including the manpower needed to install physical, hydraulic, and electric network facilities. Costs (2) and (3) refer to production costs. Variable costs refer to all those in function of production volume, which were calculated as the sum of the costs of animals, raw materials, direct manpower, industrial services and indirect production costs (IPC), which include materials, supplies, and others (Arora et al., 2018). Transportation costs were deemed as non-significant regarding the other variable costs involved, as they have been assumed in other studies on waste treatment (Bernstad et al., 2011). Fixed costs (3) were estimated as the sum of the costs that correspond to asset depreciation using the straight-line method (Weber et al., 2020), for a term of 10 years (Mota et al., 2019), as well as equipment maintenance, for which a rate of 0.01 was considered according to You et al., 2016 and Weber et al., 2020, in addition to rent, cleaning, and administrative expenses. All data on prices and costs were based on the current market value in Colombian pesos

(COP) and were converted into American dollars (USD) with a conversion rate of 3999 COP per USD (Banco de la República de Colombia, 2023).

Income

Income is the result of the sale of the products obtained on each type of farm. The methodology used was the one implemented recently by Weber et al., (2020), determining the prices in the market for 1 kilogram of fish and 1 kilogram of oil fish, whereas DCS was estimated according to the price of the kilogram of fish flour protein ($0.82/g protein), as they have similar digestibility and are a source of animal protein (Banze et al., 2017). Considering that the production cost of 1 kg of DCS was $0.54 and that the estimated sale price was $0.75), a gross profit of approximately 37% would be obtained.

Balance Points Estimation

Balance point (BP) corresponds to the production level when total income (TI) and total costs (TC) are equal (TI=TC) and there is neither profit nor loss. This parameter was estimated for each product by following the Weighted Contribution Margin (WCM), which is based on calculating the contribution margin for each product (WCMS) (Potkany et al., 2015). First, the sales participation percentage of each product (%PPs) was estimated regarding total sales (equation 1), followed by calculating the unit contribution margin (UCM) (equation 2). WCMS was calculated with equation 3, while WCM was estimated by adding WCMs. Finally, to calculate BP in total units (BPT) and BP in units for each product (BPS), equations 4 and 5 were used respectively, where TFC corresponds to total fixed costs.

$$\%PP_S = \frac{Sales_s}{Total} \tag{1}$$

$$UCM = P_{p.u} - CV_{p.u} \tag{2}$$

$$WCM_S = UCM.\%PP_S \tag{3}$$

$$BP_T = \frac{CFT}{WCM} \tag{4}$$

$$BP_s = BP_S.\%PP_s \tag{5}$$

Cost-Benefit Ratio Analysis

This analysis is based on the calculation of financial indicators, which are the result of the interaction of all the components of an investment project and are useful for assessment, decision-making, and risk identification (De Clercq et al., 2017). Such analyses were based on a factorial method using projected annual sales profit (Farid et al., 2020). Net present value (NPV) was calculated using equation 6 (You et al., 2016), in which the number of terms (LT) for which cash flow was estimated was 5 years; net cash inflow (Ct) was calculated by updating income and expenses considering an average annual inflation of 4.4% (DANE, 2022); investment costs (Co) for each system were used. The minimum acceptable rate of return (MARR), also known as the discount rate, was estimated using equation 7, where i corresponds to the premium for the risk, defined as the yield percentage that investors require. It is a function of the risk, and in this case, a medium risk of 10% was assigned. F corresponds to the average annual inflation. This rate showed similar values (~15%) to those of the discount rates reported by Manioğlu et al., (2006) for economic analyses of this type.

$$NPV = \sum_{t}^{LT} \frac{C_t}{(1+MARR)^t} - C_o \tag{6}$$

$$CBR = \frac{NPV}{\sum_{t}^{LT} \frac{C_{et}}{(1+MARR)^t} + C_o} \tag{7}$$

Additionally, the internal return rate (IRR), which refers to the profit or loss percentage that an investment or project will have, was also estimated. It corresponds to the discount rate when there is an NPV of zero. This was calculated using the algorithm provided by Matlab Software (License 914762, version 2018). The cost-benefit ratio (CBR) was determined using equation 8 (You et al., 2016), including the expenses' cost for year t (C_{et}).

$$CBR = \frac{NPV}{\sum_{t}^{LT} \frac{C_{et}}{(1+MARR)^{t}} + C_{o}} \quad (8)$$

Standard criteria were considered to analyze these parameters (Weber et al., 2020). If NPV>0, the farming model is feasible; If NPV=0, decision-making makes no difference; whereas if NPV<0 makes the model is not feasible. Meanwhile, if IRR>MARR, it is feasible; if IRR<MARR, it is not feasible; and if IRR=MARR, decision making will make no difference. For CBR, it is understood that, if CBR>0, there will be a net profit, and for a CBR<0, there will be losses (You et al., 2016).

Finally, the investment payback period (IPP) refers to the number of years required for the deducted cash flows accumulated to be equal to total costs and the initial investment (Di Trapani et al., 2014). IPP is a measure of how soon the investment can start to generate profit. Therefore, it is desirable to be as low as possible (Engelberth, 2020). This parameter was calculated using equation 9, where "A" is the last period before the investment payback, "B" is the value of the initial investment, "C" is the cash flow for period "A," and "D" is the cash flow of the period when the investment is paid back.

$$IPP = A + \left[\frac{b-c}{d}\right] \quad (9)$$

Results and Discussion

Investment Costs

Tables 1 and 2 show the costs associated with the initial investment elements for each of the processes. Based on these, total investment costs for each system assessed regarding the stages they involved were estimated, as shown in the last column of Table 2. The highest cost in all systems corresponds to fish farming facilities, which is also the main activity. The tanks require geomembranes that represent 50% of these costs. On the other hand, it can be seen that the investment cost for the FFC is lower than for the FFS because the first one corresponds just to fish production, whereas the second involves dry silage production.

Table 1. Investment costs for fish farming control and for fish silage processing plant

Fish farming control			
Concept	Quantity	Cost/u	Total cost
G1 Tank	3	728.18	2184.55
G2 Tank	4	988.25	3952.99
Hydraulic system	1	79.27	79.27
Water pumps	2	48.01	96.02
Filters	2	887.56	1775.11
pH meter	1	9.00	9.00
Oxygen sensor	1	22.26	22.26
Industrial balance	1	40.01	40.01
Baskets	6	2.25	13.50
Basket holder	1	47.51	47.51
Tables	2	47.51	95.02
Freezer	1	936.23	936.23
Fishing equipment	1	57.51	57.51
Pressure washer	1	75.02	75.02
Labor	1	187.25	187.25
Total			9571.26
Fish silage processing plant			
Gas kettle	1	500.13	500.13
Gas pipette	1	12.50	12.50
Food Processor	2	31.51	63.02
Plastic containers	4	11.00	44.01
Solar dryer	1	65.02	65.02
Traditional mill	1	50.01	50.01
Total			734.68

Table 2. Investment costs (USD) total for each system

System	Fish cost	Chemycal silage cost	Total Investment costs
FFC	9571.26	0.00	9571.26
FFS	9571.26	703.18	10274.44

Production Costs

Production costs (PC) for each farm production system are shown in Table 3. They represent the sum of fixed and variable costs. It was found that the FFC system has higher costs as it requires an external service for the final disposal of the viscera that allows it to comply with the provisions of Colombian legislation in Decree 838 of 2005 (Republic of Colombia, 2005), for a value of $0.79/kg of waste generated, thus raising the value of this concept, compared with the FFS system, which uses these subproducts to obtain CS and fish oil for commercialization, changing cost into an income. Additionally, FFC has a higher cost in terms of conventional protein raw materials such as fish flour and soy cake, which raise production costs in feed.

Table 3. Fixed and variable costs for the different production systems per cycle

Fixed costs				
Cost concept	FFC		FFS	
	USD	%	USD	%
Depreciation	1094.80	15.80	1165.40	17.88
Maintenance	109.48	1.58	116.54	1.79
Lease	0.08	1.08	75.02	1.15
Administrative	12.82	0.19	12.82	0.20
Cleanliness	57.46	0.83	57.46	0.88
Subtotal	1349.59	19.48	1427.25	21.89
Variable costs				
Feed	3532.25	50.99	3532.25	54.18
Inputs for CS	0.00	0.00	22.61	0.35
Animals	487.62	7.04	487.62	7.48
Labor	836.27	12.07	860.88	13.21
Indirect manufacturing costs	13.13	0.19	15.08	0.23
Industrial Services	708.24	10.22	173.29	2.66
Subtotal	5577.51	80.52	5091.73	78.11
Total cost	6927.10		6518.98	

The high percentage of costs that correspond to feed must be highlighted, since that item represents 63.3% and 69.4% for FFC and FFS, respectively. Those values align with others reported in fish production studies (Hadelan et al., 2012; Di Trapani et al., 2014). Costs are similar for both farms because the raw materials required to produce CS do not represent a significant increase.

Considering that facility construction costs and operational costs represent a significant part of the costs in these systems, and that the deduction rate and the other cost concepts are less controllable, strategies that allow to reduce operational and construction costs must be proposed to improve the economy of evaluated systems, with special attention to CBR values to make use of viscera by CS production for animal feed.

Income Analysis

Table 4 shows the income statement for both farm systems in a production cycle (7 months). The table shows that FFS shows higher revenue and net profits than FFC because processing viscera to produce CS and oil fish generates additional income for the FFS system. Although profit for fish sales in the FFS system is above 98%, it is noticeable that this farm model has net profits 14% higher than FFC.

Table 4. Profit and loss statement for the different production systems by cycle

Concept	FFC	FFS
Revenues (+)	11377.32	11595.26
Fish	11377.32	11377.32
Oil	--	136.17
DCS	--	81.77
Fixed costs (-)	1349.59	1427.25
Depreciation	1094.80	1165.40
Maintenance	109.48	116.54
Lease	75.02	75.02
Administrative	12.82	12.82
Cleanliness	57.46	57.46
Variable costs (-)	5577.51	5091.73
Feed	3532.25	3532.25
Inputs-CS	0.00	22.61
Animals	487.62	487.62
Labor	836.27	860.88
Indirect manufacturing costs	13.13	15.08
Industrial services	708.24	173.29
Gross profit	4450.21	5076.27
Taxes (-)	1335.06	1522.88
Net profit	3115.15	3553.39

These results evince the economic potential of using fish viscera in small and medium fish production systems, as reported by Mota et al., (2019), who designed an oil extraction system from the Nile tilapia viscera in Brazil, bringing the conclusion that coupling a system that involves technological processes to use both the protein and the lipid part of fish viscera can generate additional income and reduce the environmental impact caused by these organic macrocomponents. However, these results are not enough to determine the economic feasibility of evaluated systems. Therefore, it is necessary to make a different type of analysis, such as cost-benefit analysis (CBA).

Balance Point Analysis

Table 5 shows each product's balance points for each production system. They correspond to the units needed to be commercialized to cover total costs and not have losses in a 7-month cycle. Fish can be seen as the product that contributes the most in this aspect, showing that over 960 kg of fish need to be sold in both farming systems to reach this point. This behavior is due to the fact that the main production activity in both systems studied is fish farming. On the one hand, production volume, and on the other, because it is the starting point for the other production activities. It is important to note that the FFS systems require commercializing slightly higher fish amounts than the FFC control systems because the fixed costs of the first are higher, since they use equipment in the viscera recovery and transformation process.

Table 5. Breakeven points for the different production systems

System	Products		
	Kg Fish	Kg Oil	Kg DCS
FFC	960	--	--
FFS	988	59,3	26,1

Cost-Benefit Analysis

The technical-economic assessment using a cost-benefit analysis (CBA) is an essential tool to validate sustainability and make decisions in the farming systems studied, as it allows comparing them in profitability terms and other relevant technical-economic variables such as NPV, IRR, CBR, and PP (Table 6). Thus, NPV represents the total of the resources in favor of the farming system at the end of the period (5 years in this current study). The study showed positive values in both systems, with the FFS farm the one with the highest value in this parameter, over 20% higher than the FFC farm.

On the other hand, IRR values went over the MARR by 15.15% (You et al., 2016) for all cases, proving that both farming systems can be economically feasible. Regarding net profit and production costs, the feasibility of both systems is mostly due to fish production. However, the FFS system is favored because of oil and dry silage sales. Based on the IRR, it is concluded that the system with the highest economic revenue is FFS. A similar behavior was observed with the CBR, in which the FFS farm had a higher net revenue than the FFC farm.

Table 6. Financial evaluations for different production systems

System	NPV (USD)	IRR	CBR	IPP
FFC	9765.62	51.77	0.16	1.91
FFS	11782.77	55.88	0.19	1.33

The results obtained for the investment payback period (IPP) had a behavior aligned with NPV, IRR, and CBR results, showing to be shorter for the farm model that uses viscera to obtain oil and silage production, clearly showing a shorter payback period for FFS compared with the FFC control farm.

For the system with the best economic performance, (FFS), values of 55% and 1,3 years were obtained for IRR and IPP, respectively, which are better than those obtained by Mota *et al. (2019),* with IRR and IPP values of 45% and 1,9 years, respectively, regarding the use of Nile tilapia (*Oreochromis niloticus*) viscera to extract fish oil for biodiesel production, which are similar to those obtained in the use of sweet potato waste, which have an IRR and an IPP of 51% and 1,06 years, respectively (Weber et al., 2020). However, some studies in waste use have been considered as profitable and economically feasible despite having results below the ones in this study, such as the case of *Aspergillus niger* lipase enzyme production using agro-industrial waste with an IRR of 35% (Khootama et al., 2018), and a biorefinery of mango waste processing with an IRR of 34% (Arora et al., 2018). All these examples show that using waste, in addition to being environmentally friendly, can be economically feasible. However, there have been waste use reports in which there is no economic feasibility, such as in the case of methane production from biological waste, whose NPV was negative (De Clercq et al., 2017), or in the case of a biorefinery of algae biomass to produce biofuel and energy, which also proved to be unfeasible (Abdul and Lim, 2019). Waste use studies with unfavorable results indicate that economic feasibility in waste use is not guaranteed and that each case requires proper evaluation.

The results in the FFS model show that using fish viscera improves economic yield thanks to the sale of co-products such as oil and DCS. Currently, it is not common to find financial information on viscera use and fish waste in general in the literature, much less using the silage technique. Therefore, these results do represent a certain degree of novelty.

This is something considerable, especially when silage has been the subject of several research studies (Davies et al., 2020), and stands out as a substitute alternative for more expensive protein sources based on conventional materials such as fish flour and soy cake (Madage et al., 2015; Davies et al., 2020). It is clear that economic studies in fish systems are affected by the particular aspects of each region or country (Mota et al., 2019). However, these results may be useful for fish producers who are interested in using fish viscera with the silage technique, regardless of their region or country away from Colombia, which was the reference of this study to obtain the costs and prices.

Conclusion

The results of this study lead to the conclusion that using red tilapia (*Oreochromis* spp.) viscera derived from the production of red tilapia to obtain oil and CS improves the economic indicators of such production activity, preserving its feasibility, but under conditions that reduce waste disposal with a high contamination potential.

Acknowledgements

The authors of this document thank the Universidad de Antioquia, the Extension Program and Project University Bank of University of Antioquia (project 2020-35630), and COLCIENCIAS (project 1115-745-58746) for the financial support provided for its development.

References

Abdul, N.N., y Lim, J.S., Evaluation of processing route alternatives for accessing the integration of algae-based biorefinery with palm oil mill, https://doi.org/10.1016/j.jclepro.2018.12.104, *J. Clean. Prod.,* 212(2019), 1282–1299 (2019).

Arias, L., Gómez L.J., Zapata J.E. 2017. Efecto de temperatura-tiempo sobre los lípidos extraídos de Vísceras de tilapia roja (Oreochromis sp.) utilizando un proceso de calentamiento-congelación. *Inf Tecnol.* 28(5):131-142. doi:10.4067/s0718-07642017000500014.

Arora, A., Banerjee, J., Vijayaraghavan, R., MacFarlane, D. y Patti, AF (2018). Diseño de proceso y análisis técnico-económico de una biorrefinería integrada de residuos de procesamiento de mango. *Cultivos y productos industriales*, *116*, 24-34.

Banze, J.F., Da Silva, M.F.O., Enke, D.B.S., y Fracalossi, D.M., Acid silage of tuna viscera: Production, composition, quality and digestibility, https://doi.org/10.20950/1678-2305.2017.24.34, *Bol. Inst. Pesca,* 43, 24–34 (2017).

Bernstad, A., y Jansen, J., *A life cycle approach to the management of household food waste - A Swedish full-scale case study,* https://doi.org/10.1016/j.wasman.2011.02.026, Waste Manage, 31(8), 1879–1896 (2011).

Camaño, J.A., Rivera, A.M., y Zapata, J.E., Efecto del espesor de película y de la ubicación de la muestra en un secador solar directo, sobre la cinética de secado de ensilado de vísceras de tilapia roja (Oreochromis sp), https://doi.org/10.4067/s0718-07642020000100053, *Inf. Tecnol,* 31(1), 53–66 (2020).

Camaño, J.A., Rivera, A.M., y Zapata Montoya, J.E., *Sorption isotherms and thermodynamic properties of the dry silage of red tilapia viscera (Oreochromis spp.) obtained in a direct solar dryer,* https://doi.org/10.1016/j.heliyon.2021.e06798, Heliyon, 7(4), e06798 (2021).

DANE. (2022). Índice de precios al consumidor IPC. *Boletín Técnico.* Recuperado de: https://www.dane.gov.co/index.php/estadisticas-por-tema/precios-y-costos/indice-de-precios-al-consumidor-ipc/ipc-informacion-tecnica.

Davies, S.J., Guroy, D., Hassaan, M.S., El-Ajnaf, S.M., & El-Haroun, E. (2020). Evaluation of co-fermented apple-pomace, molasses and formic acid generated sardine based fish silages as fishmeal substitutes in diets for juvenile European sea bass (Dicentrachus labrax) production. *Aquaculture*, *521*, 735087.

De Clercq, D., Wen, Z., y Fei, F., *Economic performance evaluation of bio-waste treatment technology at the facility level,* https://doi.org/10.1016/j.resconrec.2016.09.031, Resour. Conserv. Recycl, 116, 178–184 (2017).

Di Trapani, A.M., Sgroi, F., Testa, R., y Tudisca, S., *Economic comparison between offshore and inshore aquaculture production systems of European sea bass in Italy,* https://doi.org/10.1016/j.aquaculture.2014.09.001, Aquaculture, 434, 334–339 (2014).

Engelberth, A., Evaluating Economic Potential of Food Waste Valorization: Onward to a Diverse Feedstock Biorefinery, https://doi.org/10.1016/j.cogsc.2020.100385, *Curr. Opin. Green Sustain. Chem,* 26, 100385 (2020).

FAO, *Oreochromis niloticus in cultured aquatic species fact sheets.* Text by Rakocy, J.E. edited and compiled by Valerio Crespi and Michael New. CD-ROM (multilingual), Roma, Italia (2009).

Farid, M.A.A., Roslan, A.M., Hassan, M.A., Hasan, M.Y., Othman, M.R., & Shirai, Y. (2020). Net energy and techno-economic assessment of biodiesel production from waste cooking oil using a semi-industrial plant: A Malaysia perspective. *Sustainable Energy Technologies and Assessments*, *39*, 100700.

Gaviria, Y.S., Figueroa, O.A., y Zapata, J.E., *Efecto de la inclusión de ensilado químico de vísceras de tilapia roja (Oreochromis spp.) en dietas para pollos de engorde sobre los parámetros productivos y sanguíneos,* http://dx.doi.org/10.4067/S0718-07642021000300079, Inf. Tecnol, 32(3), 79-88 (2021).

Gaviria, Y.S., Londoño, F.L.F., y Zapata, J.E., *Effects of CS of red tilapia viscera (Oreochromis spp.) as a source of protein on the productive and hematological parameters in isa-brown laying hens (Gallus gallus domesticus),* https://doi.org/10.1016/j.heliyon.2020.e05831, Heliyon, 6(12) (2020).

Gonzales, C., y Quevedo, E., *Cultivo de las tilapias roja (Oreochromis spp.) y plateada (Oreochromis nilotus)* (p. 17), Pereira, Colombia (2018).

Gomez, G.M., Ortiz, M.A., Perea, C., & Lopez, F.J. (2014). Evaluación del ensilaje de vísceras de tilapia roja (Oreochromis spp) en alimentación de pollos de engorde. *Biotecnología en el sector agropecuario y agroindustrial*, *12*(1), 106-114.

Goosen N.J., de Wet L.F., Görgens J.F., Jacobs K., De Bruyn A. 2014. Fish silage oil from rainbow trout processing waste as alternative to conventional fish oil in formulated diets for Mozambique tilapia Oreochromis mossambicus. *Anim Feed Sci Technol.* 188:74-84. doi:10.1016/j.anifeedsci. 2013.10.019. http://dx.doi.org/10.1016/j.anifeedsci.2013.10.019.

Güllü K., Acar Ü., Tezel R., Yozukmaz A. 2014. Replacement of fish meal with fish processing by-product silage in diets for the rainbow trout, Oncorhynchus mykiss. *Pak J Zool.* 46(6):1697- 1703.

Hadelan, L., Par, V., Njavro, M., y Lovrinov, M., Real option approach to economic analysis of European sea bass (Dicetrarchus labrax) farming in Croatia, *Agric. Conspec. Sci,* 77(3), 161–165 (2012).

Khootama, A., Putri, D.N., y Hermansyah, H., *Techno-economic analysis of lipase enzyme production from Aspergillus Niger using agro-industrial waste by solid state fermentation,* https://doi.org/10.1016/j.egypro.2018. 10.054, Energy Procedia, 153, 143–148 (2018).

Madage, S.S. K., Medis, W.U.D., y Sultanbawa, Y., *Fish silage as replacement of fishmeal in red tilapia feeds,* https://doi.org/10.1080/10454438.2015. 1005483, *J. Appl. Aquac,* 27(2), 95–106. (2015).

Manioğlu, G., y Yilmaz, Z., Economic evaluation of the building envelope and operation period of heating system in terms of thermal comfort, https://doi.org/10.1016/j.enbuild.2005.06.009, *Energy Build,* 38(3), 266–272 (2006).

Mota, F.A.S., Costa, J.T., y Barreto, G.A., The Nile tilapia viscera oil extraction for biodiesel production in Brazil: An economic analysis, https://doi.org/ 10.1016/j.rser.2019.03.035, *Renew. Sust. Energ. Rev,* 108(March), 1–10 (2019).

Potkany, M., y Krajcirova, L., *Quantification of the volume of products to achieve the break-even point and desired profit in non-homogeneous production,* https://doi.org/10.1016/s2212-5671(15)00811-4, Procedia Econ. Financ, 26(15), 194–201 (2015).

República de Colombia, Decreto 838 de 2005, *Por el cual se modifica el Decreto 1713 de 2002 sobre disposición final de residuos sólidos y se dictan otras disposiciones,* Bogotá, Colombia (2005).

Suarez, L.M., Montes, J.R., y Zapata, J.E., *Optimización del contenido de ácidos en ensilados de vísceras de Tilapia roja (Oreochromis spp.) con análisis del ciclo de Vida de los alimentos derivados,* https://doi.org/10.4067/s0718-07642018000600083, *Inf. Tecnol.,* 29(6), 83–94 (2018).

van't Land, M. y Raes, K. (2019). Secado de ventana de refractancia del ensilaje de pescado: una investigación inicial sobre los efectos de las propiedades fisicoquímicas en la eficiencia del secado y la calidad nutricional. *LWT, 102,* 71-74.

Villamil, O., Váquiro, H., y Solanilla, J.F., Fish viscera protein hydrolysates: Production, potential applications and functional and bioactive properties, https://doi.org/10.1016/j.foodchem.2016.12.057, *Food Chem,* 224, 160–171 (2017).

Weber, C.T., Trierweiler, L.F., y Trierweiler, J.O., Food waste biorefinery advocating circular economy: Bioethanol and distilled beverage from sweet potato, https://doi.org/10.1016/j.jclepro.2020.121788, *J. Clean. Prod,* 268 (2020).

You, S., Wang, W., Dai, Y., Tong, Y.W., & Wang, C.H. (2016). Comparison of the co-gasification of sewage sludge and food wastes and cost-benefit analysis of gasification-and incineration-based waste treatment schemes. *Bioresource technology*, *218*, 595-605.

Chapter 5

Ecological Footprint Methodology as an Indicator of Sustainability in Alternative Feeding of Broiler Chickens

Yhoan S. Gaviria[1]
Omar A. Figueroa[2], **PhD**
and José E. Zapata[1,*], **PhD**

[1]Department of Food and Pharmaceutical Sciences, Food Nutrition and Technology Group, Universidad de Antioquia, Medellín, Colombia
[2]Faculty of Engineering and Basic Sciences. Fundación Universitaria del Área Andina, Valledupar, Colombia

Abstract

Globally, the fish farming industry has one of the highest growth rates in recent decades, however, this causes a massive production of viscera, scales, skeletons, among other by-products. The valorization of fish biowaste as a feedstock to formulate foods used in the production chain of this industry is a research key area to achieve the Sustainable Development Goals (SDGs). However, the environmental impact of the use of this waste in that sense is not known yet. In the present study, it was proposed to determine the environmental impact of the use of chemical silage of red tilapia (*Oreochromis* spp.) viscera as a source of protein in the preparation of food for broilers (*Gallus gallus domesticus*) of the Ross 308 line, which have demonstrated their effectiveness without altering the productive variables of the foods

[*] Corresponding Author's E-mail: edgar.zapata@udea.edu.co.

In: Agro-Industrial Wastes
Editor: Peter Clements
ISBN: 979-8-89113-688-5

designed. After obtaining chemical silage from red tilapia viscera, these were implemented as a protein substitute in chicken production, with monitoring of zootechnical variables for 45 days. Biosafety and health conditions were followed according to National Federation of poultry farmers of Colombia (FENAVI) recommendations. Finally, the environmental impact of the production system was quantified using the ecological footprint methodology as an indicator of sustainability in the feeding of broilers. The implementation of silage in the feeding of broilers did not have a negative effect on the productive variables, weight gain, size, or conversion, which, in addition, were in the normal range for this species. Additionally, silage production generated a 15% reduction in the environmental impact on the production system, corresponding to -0.14 hectares of forest per cycle.

Keywords: chemical silage, *Oreochromis* spp, Isa Brown, investment cost

Introduction

Economic and productive development worldwide creates ecological pressure on the environment as it generates a high demand on resources such as water, food, infrastructure, and energy, which translates into the reduction of natural resources, the accumulation of waste, and the emission of greenhouse gases (Ahmed & Wang, 2019). Fish farming is one of the industries with the highest growth rate in the last decades, reaching 171 million tons worldwide in 2016 (FAO, 2018). Consequently, this industry represents one of the highest-impact productive activities, since up to 65% of its production becomes waste (Martínez-Alvarez et al., 2015). Some of this waste, like entrails, can be used as a protein source for the animal feed industry. Therefore, entrails silage has been implemented in different diets for animal species, giving added value to this waste, extending conservation time, and improving nutritional composition (Gaviria G et al., 2020; Gomez et al., 2014; Suarez et al., 2018).

On the other hand, in the poultry industry, 95% of the total cost of the diets is used to supply protein and energy needs. In addition, the poultry industry has been the fastest-growing livestock sector recently, mostly driven by strong demand (Farrell, 2013). The main protein sources in this industry are soy cake and corn flour, respectively (Ravindran, 2013). However, because of its high demand, acquisition costs are high, and countries like Colombia have to import, which increases production costs even more for poultry feed, which has led to a search for low-cost protein alternatives (Ravindran, 2013). This has brought fish farming waste silage as a protein source in the feed of species such as laying hens, Japanese quail and Mozambique tilapia, with positive results (Gaviria G et al., 2020; Goosen et al., 2014; Ramírez et al., 2013). Still, there are no literature reports that quantify the environmental impact of their use for these types of species.

The ecological footprint has been one of the most commonly used process sustainability indicators to assess environmental impact (Gwehenberger & Narodoslawsky, 2007). It quantifies the effect of anthropogenic activities on the environment regarding soil, water, forest products, infrastructure, and carbon footprint, providing comparable, reliable, and comprehensive responses (Ahmed & Wang, 2019; Jóhannesson et al., 2018). This study is intended to determine the environmental impact of using red tilapia (*Orcochromis* spp.) entrails to obtain chemical silage and its implementation for feeding broiler chickens of the Ross 308 line, using the methodology of ecological footprint as a sustainability indicator.

Methodology

The production system consisted of a red tilapia (*Oreochromis* spp.) fish farm located in the municipality of San Jerónimo, in Antioquia – Colombia (6°26′30″N 75°43′40″O), which uses entrails waste to obtain chemical silage, and implements it in diets for feeding broiler chickens. Figure 1 shows the sequence of the processes carried out in the production system, indicating the outcome of products and waste in each one of them.

Subsystem 1 (Subs1): Makes reference to red tilapia (*Oreochromis* spp.) breeding and fattening stages, in which, 2044 fingerlings were fed for 7 months until they reached an average weight of 400 g. Fish were fed with three diets of commercial feed according to their growth stage.

Subsystem 2 (Subs2): This stage involves fish processing, which is when they are also eviscerated. A total of 1778 fish were processed in this stage, with a death rate of 13%, and entrails waste represented 16% of the total weight obtained (711 kg).

Subsystem 3 (Subs3): Corresponds to the chemical silage (CS) manufacture process from red tilapia entrails, which is carried out in several stages: degreasing, grinding, silaging, and storing, as reported by Suarez et al., (2018). 100% of the chemical silage obtained was implemented in feed for broiler chickens.

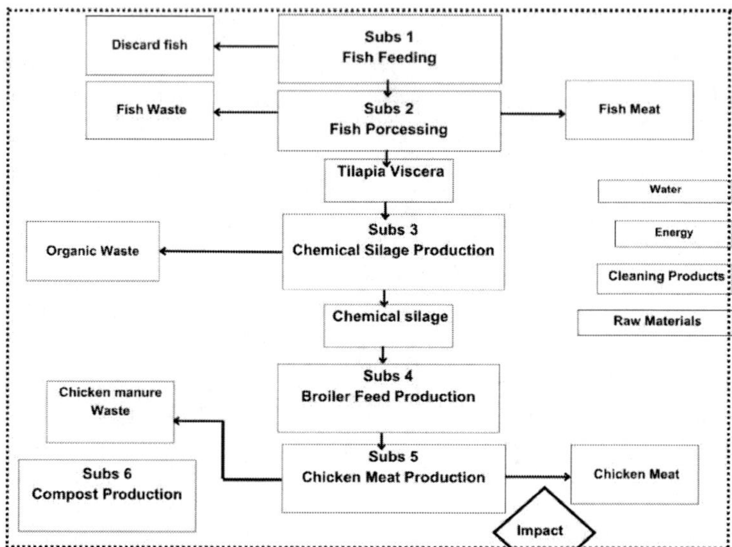

Figure 1. Limit of the production system process. Source: authors.

Table 1. Experimental diet for broilers

Source	Silage diet (%)
Cornmeal	40
Rice flour	17
Silage	25.5
Soy cake	9
Fish flour	5
Fish oil	0
Calcium carbonate	1
Calcium phosphate	1
Vitamin Supplement	0.50
Lysine	0.25
Methionine	0.25
Tryptophan	0.25
Threonine	0.25
Protein (%)	21.10
Carbohydrates (%)	41.00
Fat (%)	7.4
Metabolizable energy (Kcal/kg)	3500
Calcium (%)	1.10
Phosphor (%)	0.60

* Composition per 250 g of product: vit. A – 1,400,000 IU; vit. B1 - 500 mg; vit. B12 - 300 mg; vit. B2 = 500 mg; vit. B6 – 1,6 g; vit. D3 – 2,500,000 IU; vit. E – 6.000 IU; vit. K3 = 1,000 mg; biotin - 30 mg; niacin -12 g; folic acid - 1 g; cobalt - 50 mg; copper – 3.000 mg; iron - 25 g; iodine - 500 mg; manganese - 32.5 g; selenium - 100.50 mg; zinc - 22.49 g.

Subsystem 4 (Subs4): Consists of the manufacture of feed for broiler chickens, which involves 3 stages: mixing, pelletizing, and drying. Table 1 shows the defined formulation and characterization of the feed for poultry aligned with their nutritional requirements, according to the Brazilian nutrition tables (Rostagno et al., 2011).

Subsystem 5 (Subs5): Corresponds to obtaining chicken meat, for which 100 Ross 308 broiler chickens (*Gallus gallus domesticus*) were fed from 0 to 45 days of age. Chickens were fed twice a day (morning and afternoon) with a ration that was increased progressively each week, in addition to a constant ad libitum water supply.

Subsystem 6 (Subs6): Makes reference to composting the excretions waste from the feeding stages of the birds for a period of 30 days.

Assumptions and Limitations

Some assumptions were considered for this study, such as:

1) The CS manufacture process was carried out in the same farm where fish are bred, to eliminate the effect of transporting the entrails.
2) The impact of buildings was not considered in the calculations as their lifespan is extended.
3) The environmental impact of industrial equipment was not included within the environmental analysis. However, it has been proven that these do not significantly affect the results in the environmental impact assessment as they have long lifespan periods (Perez-Martinez et al., 2018).
4) In Subs4, it is considered that the water steam molecules are the only compounds released into the environment due to the drying process, given that the system does not reach evaporation temperatures for other compounds.

Environmental Impact Assessment

The ecological footprint methodology was implemented to determine the environmental impact present in the production system observed in Figure 1. This methodology was used to determine the environmental impact within the production system observed in Figure 1. It was carried out as proposed by Gwehenberger & Narodoslawsky in 2007, and by Krotscheck & Narodoslawsky in 1996, hence estimating the ecological impact of each stage using calculation methods.

A number of, 2044 red tilapia (*Oreochromis* spp.) fish bred were considered the fundamental unit to quantify the ecological impact. The values of the environmental impact of the input and output in each stage of the system were normalized to the same unit (Ha/ton), which represents the amount of forest hectares needed in the specified region to fixate CO_2 produced during each stage.

To calculate the environmental impact caused by the organic compounds (IAco) required or generated during the different process stages, equation 1 was used implementing the directives from the Intergovernmental Panel on Climate Change (IPCC), where CRO corresponds to the amount of organic waste, FCM is the correction factor for methane gas (CH_4), which depends on the process associated with solid waste management in a particular sector (IPCC, 2006), COD represents the degradable organic carbon fraction that can be subject to biochemical decomposition (IPCC, 2006), COD_F is the non-assimilated COD fraction or that degrades very slowly (IPCC, 2006), F and R correspond to the CH_4 fraction in the dumping gas and the CH_4 fraction recovered, respectively. The CH_4 oxidation factor is represented as OX, methane PCG global warming potential was used for a 100-year period (IPCC, 2006) and FF corresponds to the CO_2 fixation factor for the region established within the environmental analysis (IDEAM, 2011).

$$IA=\sum_{co=1}^{n} RO\left(\frac{16*(FM*COD*COD_F*(F-R)*(1-OX)*PCG)}{12*FF}\right) \quad (1)$$

Equation 2 was implemented to calculate the environmental impact caused by the input of electric energy in the process stages that require it, where EE_G corresponds to the electric energy expense “e” in each stage of the process, and EEE is the CO_2 emission factor caused by one kw-h of energy (IDEAM, 2011).

$$IA=\sum_{e=1}^{n} EE_G*\frac{EEE}{FF} \quad (2)$$

During the chemical silage manufacture process (subs 3), it is necessary to heat entrails with propane gas to degrease them. Equation 3 is used to determine its impact, where VR_{C3H8} is the gas volume required, EBp is the energy imbued in the process to obtain propane gas, and D_R, R_c, and C are the traveled distance in transportation, fuel efficiency, and the propane gas load, respectively. Finally, CO_2 emissions from the fuel were defined as EC. Additionally, this system generates CO_2 from the combustion of propane gas, which was calculated using Equation 4, where FC_{C3H8} corresponds to the conversion factor for propane gas, and FE_{C3H8} is the CO_2 emission factor (IPCC, 2006).

$$IA=VR_{C_3H_8}\left(\frac{EB_P*EEE+\left(\frac{D_R}{R_C*C}\right)*EC}{FF}\right) \quad (3)$$

$$IA=\frac{VR_{C_3H_8}*FC_{C_3H_8}*FE_{C_3H_8}}{FF} \quad (4)$$

The impact caused by water supply was calculated using Equation 5, where A_L corresponds to the water liters required in the process (a) and EB_A is the energy imbued for water supply. However, using it implies an environmental impact associated with its final disposal, which is calculated using Equation 6, where ARL is the waste water obtained in the cleaning process, CO_{DBO5} is the oxygen biochemical demand of the waste water degradable fraction, and CMP_{CH4} indicates the maximum capacity of methane production of that fraction (IPCC, 2006).

$$IA=\sum_{a=1}^{n} A_L\left(\frac{EB_A*EEE}{FF}\right) \quad (5)$$

$$IA=\sum_{a=1}^{n} AR_L*\left(\frac{CO_{DBO5}*CMP_{CH4}*FCM_{CH4}*PCG}{FF}\right) \quad (6)$$

Equation 7 is used for the environmental impacts associated with the production ingredients and any other input flows in the different subsystems, where CR_I is the amount of product "I" required, EB_{PI} corresponds to the energy imbued to manufacture it, and C_I the maximum load for product transportation (IPCC, 2006).

$$IA=CR_I\left(\frac{EB_{PI}*EEE+\left(\frac{D_R}{R_C*C_I}\right)*EC}{FF}\right) \quad (7)$$

Productive Variables

Productive variables of broiler chickens were registered weekly until they reached 45 days of age. The weekly weight of the animals was evaluated using an analytical, TxB220-1L, 1g precision scale (Shimadzu, Japan). The size of the birds was determined using a vernier, measuring the tarsus-metatarsus bone. The food conversion index (FCI) was calculated as the relation between the feed consumed and the weight gain of every bird. Finally, the death rate percentage was determined for both diets.

Results and Discussion

Productive Variables

Table 2 shows the productive variables of broiler chickens fed with a diet made with CS from red tilapia entrails. It can be observed how weight constantly increases toward the final week, showing weights within the typical range established for the Ross 308 poultry line (Rosero et al., 2012). Boitai et al. (2018) and Venturoso et al. (2016) obtained weights of 1701 and 1660 grams, respectively, in broiler chicken fed with concentrates with a partial substitution of the conventional protein using chemical silage from fresh water fish entrails and gills, reporting that the substitution did not have any adverse effects on the birds' weight. Regarding size, there is a similar tendency, showing a constant increase during the 7 weeks of feeding, and having similar results to those reported by Andrade-Yucailla et al. (2015) and Lázaro et al. (2012) in the measurement of the tarsus-metatarsus bone of hens (*Gallus gallus domesticus*), with values between 10,47 and 14,14 cm for birds with average weights of 1603 – 1920 grams for both studies.

Table 2. Broiler chicken productive variables

Weeks	Weight (g)	Size (cm)	Food consumption (g/ bird day)	Mortality (%)	Conversion Rate
0 - 1	43.06 ± 3.42	5.10 ± 0.15	16 ± 1.64	0	1.36 ± 0.05
1 - 2	111.84 ± 6.37	6.11 ± 0.18	44 ± 3.51	0	1.64 ± 0.12
2 - 3	291.47 ± 5.78	8.29 ± 0.27	70 ± 2.63	3	1.74 ± 0.09
3 - 4	562.58 ± 8.41	9.17 ± 0.32	112 ± 2.92	3	1.87 ± 0.31
4 - 5	874.00 ± 10.94	11.30 ± 0.35	135 ± 3.72	0	1.69 ± 0.23
5 - 6	1360.15 ± 14.24	12.86 ± 0.58	156 ± 3.39	0	1.78 ± 0.23
6 - 7	1830.84 ± 27.68	13.79 ± 0.66	179 ± 5.81	0	1.72 ± 0.31

Food consumption was carried out in alignment with the nutritional requirements of the species, showing progressive consumption throughout the 45 days of the study. Similar results were published by Al-Marzooqi et al. (2010), which evaluated the effect of different inclusion levels of sardine chemical silage in broiler chickens feed, finding that food consumption is positively affected by increasing silage. Regarding the death rate, an accumulated value of 6% was obtained throughout the 7 feeding weeks. This value is higher than the one reported by Shabani et al. (2019), who included biological silage from fish waste in diets for broiler chicken feed with death rate percentages within a range of 2,08–2,96%. However, compared with the values reported by Venturoso et al. (2016) in broiler chicken feed with acid silage from fish waste, it is observed that these are lower, since death rate percentages of 60–90% were found due to the formation of biogenic amines that acted on the intestinal microbiota of birds, modifying and promoting a septicemia framework that evolved and led to the reported death rate.

Food conversion index (FCI) stayed within the normal range established for the poultry line used (Ross 308) (Rosero et al., 2012), which indicated that the inclusion of chemical silage from red tilapia entrails in broiler chickens feed does not adversely impact the food conversion index and its palatability. Similar results were reported by Boitai et al., (2018) and Garcés et al., (2015), using chemical silage from fresh water fish entrails and also using rainbow trout entrails in broiler chickens feed (Santana-Delgado et al., 2008).

Quantification of the Environmental Impact of Using Chemical Silage in Poultry Feed

Table 3 shows the environmental impact on ha/cycle of each stage of the production system (Figure 1) as well as the amount of the main product generated in each of them. In fish farming (subs 1), there was an average reduction of 13% of the initial population (2044 fish) because of mortality. This is the typical range of a production cycle of this nature, which is between 10 and 20% (Alicorp, 2010). This subsystem has the highest environmental system of the production process (0,57 ha/cycle) due to the volume of food consumed by fish throughout their growth, in addition to some raw materials used for fulfilling this requirement, which are associated with high emissions of greenhouse gases (Henriksson et al., 2015).

Fish processing (Subs 2), begins with removing the scales (5%) and entrails (16%), which generates an average yield of 79% in weight of the total product. There is a considerably higher environmental impact in this stage compared to previous stages because of the high volume of water required during the evisceration process, in addition to the resulting organic load in waste water. This has a direct impact due to its final disposal (Quiroz Fernández et al., 2018).

The process to obtain chemical silage in subsystem 3 has a yield of 71% in the degreasing stage. It produces 80,73 kg of CS a month from the waste of the previous stage. Then, it is used in the process of manufacturing feed as a protein substitute for conventional raw materials (soy cake and fish flour) for broiler chickens (Ravindran, 2013). This process has an environmental impact value of -0,14 ha/month. Its negative sign means that it uses the subproducts (entrails). If it did not, there would be an adverse environmental impact equivalent to that magnitude, but with a positive sign. Similar behaviors were reported by Malakahmad et al., (2017), who determined the environmental impact of several organic waste disposal processes. They found out that CO2 emissions avoided in anaerobic digestion processes are considerably higher than the total CO2 emissions of the process, which results in an expression with a negative sign, meaning that the process is favoring the environment.

Subsystem 4 corresponds to the process of broiler chicken feed manufacture, which has a relatively lower environmental impact compared with the other stages of the production system. Its main impact categories focus on the obtaining of raw materials and the use of electric energy needed for the concentrate pelletization process. Feed was dried using solar energy through solar collectors, as reported by Camaño et al., (2020), aiming to reduce the environmental load from the feed manufacturing stage. This leads the process to be framed within the tendency of the use of renewable energies in feed drying, since it has been proven in the environmental performance evaluation of various farms and processing plants that one of the main environmental pressure points focuses on the obtaining of feed because of the high consumption of electric and caloric power, and the use of traditional raw materials such as soy seed, which require the administration of mineral fertilizers (Skunca et al., 2018).

A productive variable that presents an important relevance for the environmental aspect in the production processes of the poultry industry is the food conversion index (FCI), since species such as broiler chickens emit a significant amount of greenhouse gases during their digestive process, like methane gas. This variable relates the quantity of gases produced per unit of feed consumed to the amount of meat produced (Ibid hi et al., 2017).

Table 3. Environmental impact of the production system

Subsystem	Amount	IA (Ha/cycle)
Subs 1 (unid)	1778	0.57
Subs 2 (kg)	561.94	0.47
Subs 3 (kg)	80.73	-0.14
Subs 4 (kg)	351.56	0.09
Subs 6 (kg)	187.00	0.26
Subs 7 (kg)	131.20	-0.07
Total		1.18

In subsystem 5, the values of the environmental impact correspond to the chickens' breeding and fattening stages. It is mainly affected by birds' food consumption, as it is directly related to manure production and its composition, which influences the emission of the poultry production system (growth and product obtaining). Similar tendencies were published by Leinonen et al., (2012), who reported that changes in consumption and feed composition have effects both in the environmental impacts that take place during farming production and food processing, as well as in the subsequent emissions of poultry manure during housing, storage, and field application.

The manure obtained from poultry species has physical and chemical properties that provide it with qualities that allow it to be used either as fertilizer or feed for some animal species (Perez López et al., 2019). However, when manure is not used or properly disposed, it is an important source of direct emissions into the environment of gases such as ammonia (NH3), nitrous oxide (N2O), and methane, which are produced during the fattening, storing, and land spreading period (Leinonen et al., 2012). In subsystem 6, manure produced has a negative value, this means that it was a credit instead of an environmental load because of its implementation as organic fertilizer, which compensated synthetic fertilizer production or crop rotation that fixate nitrogen (Sinha et al., 2010).

Figure 2 shows representation percentages of the main categories of environmental impact of the production system inputs and outputs. The electric energy category is characterized as one of the environmental indicators with the highest impact due to the load that obtaining it involves (Tian et al., 2012). Solar energy was implemented for the broiler chickens' concentrate drying process. It is considered as a renewable, abundant, and free energy, and it led to a significant reduction in the environmental impact and gas emissions in this category, which made it one of the most promising alternatives to reduce negative environmental effects from the drying processes (Tiwari, 2016).

Regarding the process to obtain water and its final disposal (waste water), this last category has the highest negative environmental impact from the production system (55%). That occurs mainly due to the organic load it has because the environment requires a significant amount of oxygen to degrade for its assimilation, which reduces its capacity to absorb contaminating loads and naturally restores its quality (Quiroz Fernández et al., 2018). The natural resources necessary to grow raw materials and supplies to produce fish and poultry feed represent the source with the second-highest environmental impact of the production system, with a 42,76%, having soy as one of the main greenhouse gas enhancers (GWP100) (Leinonen et al., 2012). The use of fish (entrails) and poultry (manure) industry subproducts is directly beneficial for the environment, as they prevent the generation of greenhouse gas emissions. Additionally, it gives added value to the subproducts, which has an effect on the economy of the process.

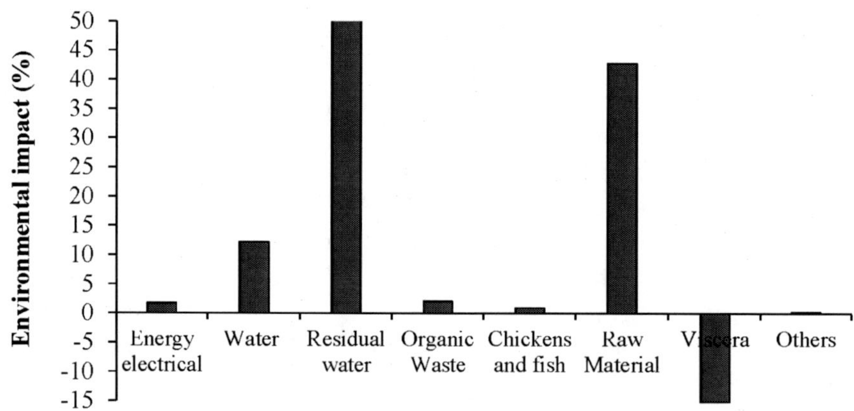

Figure 2. Environmental impact categories of the production system. Source: authors.

Conclusion

The productive variables of broiler chickens bred dieting with a partial substitution of protein for chemical silage from red tilapia entrails are not negatively affected. That is an indicator that chemical silage can be used as an alternative raw material in poultry feed to obtain meat.

The use of fish and poultry industry subproducts is the main reducer of adverse effects, such as those caused by greenhouse gas enhancers. Additionally, this strategy generates products with added value. Likewise, using alternative energies such as solar energy significantly reduces the environmental load of drying processes because, in addition to being renewable, it is also highly efficient. On the other hand, the use of natural resources and waste water treatment are the main sources of adverse effects on the environment, as they promote greenhouse gas enhancers and the oxygen demand they require, respectively.

Acknowledgements

The authors thank the Committee of Research Development of the Universidad de Antioquia (Comité de Desarrollo de la Investigación de la Universidad de Antioquia – CODI) through the sustainability program, and COLCIENCIAS (Project Code 111574558746). The authors also thank the Livestock Technological Development Center of Cesar (CDT Ganadero del Cesar).

Disclaimer

None

References

Ahmed, Z., & Wang, Z. Investigating the impact of human capital on the ecological footprint in India: An empirical analysis. *Environmental Science and Pollution Research* (2019) 26(26), 26782–26796.

Al-Marzooqi, W., Al-Farsi, M. A., Kadim, I. T., Mahgoub, O., & Goddard, J. S. The effect of feeding different levels of sardine fish silage on broiler performance, meat quality and sesory characteristics under closed and open-sided housing systems. *Asian-Australasian Journal of Animal Sciences* (2010) 23(12), 1614–1625.

Andrade-Yucailla, V., Vargas-Burgos, J. C., Lima-Orozco, R., Andino, M., Quiteros, R., & Torres, A. Caracterización morfométrica y morfológia de la gallina criolla (gallus domesticus) del cantón Carlos Julio Arosemena, Ecuador. *Actas Iberoamericanas de Conservación Animal* (2015) 6(June 2016), 42–48.

Boitai, S. S., Babu, L. K., Pati, P. K., Pradhan, C. R., Tanuja, S., Kumar, A., & Panda, A. K. Effect of dietary incorporation of fish silage on growth performance, serum biochemical parameters and carcass characteristics of broiler chicken. *Indian Journal of Animal Research* (2018) 52(7), 1005–1009. https://doi.org/10.18805/ijar.B-3315

Camaño, J. A., Rivera, A. M., & Zapata, J. E. Efecto del espesor de película y de la ubicación de la muestra en un secador solar directo, sobre la cinética de secado de ensilado de vísceras de tilapia roja (*Oreochromis* sp). *Información Tecnológica* (2020) 31(1), 53–66.

FAO. (2018). *The State of world fisheries and aquaculture. Meeting the Sustainable Development Goals.* https://doi.org/Licencia: CC BY-NC-SA 3.0 IGO

Farrell, D. (2013). *Función de las aves de corral en la nutrición humana.* Fao, 2.

Garcés, Y., Perea, C., Valencia, N. F., Hoyos, J. L., & Gómez, J. A. Efecto nutricional del ensilado químico de subproductos piscícolas en la alimentación de pollos de engorde (Gallus domesticuset al). *Cuban Journal of Agricultural Science* (2015) 49(4), 503–508.

Gaviria G, Y. S., Londoño F, L. F., & Zapata M, J. E. Effects of chemical silage of red tilapia viscera (*Oreochromis* spp.) as a source of protein on the productive and hematological parameters in isa-brown laying hens (*Gallus gallus domesticus*). *Heliyon* (2020) 6(12).

Gomez, G. M., Ortiz, M. a, Perea R, C., & Lopez, F. J. Evaluación del ensilaje de vísceras de tilapia roja (*Oreochromis* spp) en alimentación de pollos de engorde. *Biotecnología En El Sector Agropecuario y Agroindustrial* (2014) 12(1), 106–114.

Goosen, N. J., de Wet, L. F., Görgens, J. F., Jacobs, K., & de Bruyn, A. Fish silage oil from rainbow trout processing waste as alternative to conventional fish oil in formulated diets for Mozambique tilapia *Oreochromis* mossambicus. *Animal Feed Science and Technology* (2014) 188, 74–84.

Gwehenberger, G., & Narodoslawsky, M. The ecological impact of the sugar sector- Aspects of the change of a key industrial sector in Europe. *Computer Aided Chemical Engineering* (2007). 24, 1029–1034.

Henriksson, P. J. G., Rico, A., Zhang, W., Ahmad-Al-Nahid, S., Newton, R., Phan, L. T., Zhang, Z., Jaithiang, J., Dao, H. M., Phu, T. M., Little, D. C., Murray, F. J., Satapornvanit, K., Liu, L., Liu, Q., Haque, M. M.Kruijssen, F., De Snoo, G. R., Heijungs, R., … Guinée, J. B. *Comparison of Asian*

Aquaculture Products by Use of Statistically Supported Life Cycle Assessment. Environmental Science and Technology (2015) 49(24), 14176–14183.

Ibidhi, R., Hoekstra, A. Y., Gerbens-Leenes, P. W., & Chouchane, H. Water, land and carbon footprints of sheep and chicken meat produced in Tunisia under different farming systems. *Ecological Indicators* (2017) 77, 304–313.

IDEAM. (2011). *Estimación de las emisiones de dióxido de carbono generadas por deforestación durante el periodo 2005-2010.*

IPCC. (2006). *Directrices para los inventarios nacionales de gases de efecto invernadero - Elimimación de desechos sólidos - Capitulo 3.* In IPCC.

Jóhannesson, S. E., Davíðsdóttir, B., & Heinonen, J. T. Standard Ecological Footprint Method for Small, Highly Specialized Economies. Ecological Economics (2018) 146(December 2016), 370–380.

Krotscheck, C., & Narodoslawsky, M. The Sustainable Process Index. A new dimension in ecological evaluation. *Ecological Engineering* (1996) 6(4), 241–258.

Lázaro, G. C., Hernández Z, J. S., Vargas L, S., Martínez, L. A., & Pérez, A. R. Uso De Caracteres Morfometricos En La Clasificación De Gallinas Locales. *Actas Iberoamericanas de Conservación Animal* (2012) 2(November), 109–114.

Leinonen, I., Williams, A. G., Wiseman, J., Guy, J., & Kyriazakis, I. Predicting the environmental impacts of chicken systems in the united kingdom through a life cycle assessment: Egg production systems. *Poultry Science* (2012) 91(1), 26–40.

Malakahmad, A., Abualqumboz, M. S., Kutty, S. R. M., & Abunama, T. J. Assessment of carbon footprint emissions and environmental concerns of solid waste treatment and disposal techniques; case study of Malaysia. *Waste Management* (2017) 70, 282–292.

Martínez-Alvarez, O., Chamorro, S., & Brenes, A. Protein hydrolysates from animal processing by-products as a source of bioactive molecules with interest in animal feeding: A review. *Food Research International* (2015) 73(1069), 204–212.

Perez-Martinez, M. M., Noguerol, R., Casales, B. I., Lois, R., & Soto, B. Evaluation of environmental impact of two ready-to-eat canned meat products using Life Cycle Assessment. *Journal of Food Engineering* (2018) 237(May), 118–127.

Perez López, D. D. J., Monroy, J. P., Reyes Ramírez, A. K., Huerta, A. G., & Sangermán Jarquín, D. M. Fertilización orgánica con tres niveles de gallinaza en cuatro cultivares de papa. *Revista Mexicana de Ciencias Agrícolas* (2019). 10(5), 1139–1149.

Quiroz Fernández, L. S., Izaquierod Kulich, E., & Menéndez Gutierrez, C. Estudio del impacto ambiental del vertimiento de aguas residuales sobre la capacidad de autodepuración del río Portoviejo, Ecuador. *Centro Azúcar* (2018) 45(01), 73–83.

Ramírez, J. C. R., Ibarra, J. I., Romero, F. A., Ulloa, P. R., Ulloa, J. A., Matsumoto, K. S., Cordoba, B. V., & Manzano, M. ángel M. Preparation of biological fish silage and its effect on the performance and meat quality characteristics of quails (Coturnix coturnix japonica). *Brazilian Archives of Biology and Technology* (2013) 56(6).

Ravindran, V. Principales ingredientes utilizados en las formulaciones de alimentos para aves de corral Velmurugu. *Fao* (2013). 3.

Rosero, J., Ferney, E., & Lopez, F. Evaluación Del Comportamiento Productivo De Las Lineas De Pollos De Engorde Cobb 500 Y Ross 308. *Biotecnología En El Sector Agropecuario y* Agroindustrial (2012) 10(1), 8–15.

Rostagno, H. S., Texeira, L. F., Lopez, J., Cezar, P., Flávila de Oliveira, R., Clementino, D., Soares, A., Lluiz de Toledo, S., & Frederico, R. *Tablas Brasileñas Para Aves Y Cerdos* (2011) 157–166.

Santana-Delgado, H., Avila, E., & Sotelo, A. Preparation of silage from Spanish mackerel (Scomberomorus maculatus) and its evaluation in broiler diets. *Animal Feed Science and Technology* (2008) 141(1–2), 129–140.

Shabani, A., Jazi, V., Ashayerizadeh, A., & Barekatain, R. Inclusion of fish waste silage in broiler diets affects gut microflora, cecal short-chain fatty acids, digestive enzyme activity, nutrient digestibility, and excreta gas emission. *Poultry Science* (2019) 98(10), 4909–4918.

Sinha, R. K., Valani, D., Chauhan, K., & Agarwal, S. Embarking on a second green revolution for sustainable agriculture by vermiculture biotechnology using earthworms: Reviving the dreams of Sir Charles Darwin. *Journal of Agricultural Biotechnology and Sustainable Development* (2010) 2(7), 113–128.

Skunca, D., Tomasevic, I., Nastasijevic, I., Tomovic, V., & Djekic, I. Life cycle assessment of the chicken meat chain. *Journal of Cleaner Production* (2018) 184, 440–450.

Suarez, L. M., Montes, J. R., & Zapata, J. E. Optimización del Contenido de Ácidos en Ensilados de Vísceras de Tilapia Roja (*Oreochromis* spp.) con Análisis del Ciclo de Vida de los Alimentos Derivados. *Información Tecnológica* (2018) 29(6), 83–94.

Tian, M., Gao, J., Zheng, Z., & Yang, Z. The Study on the Ecological Footprint of Rural Solid Waste Disposal-example in Yuhong District of Shenyang. *Procedia Environmental Sciences* (2012) 16(0), 95–101.

Tiwari, A. A Review on Solar Drying of Agricultural Produce. *Journal of Food Processing & Technology* (2016) 7(9).

Venturoso, O. J., Reinicke, F., Da Silva, C. C., Vieira, E. O., Porto, M. O., Cavali, J., Tortato, N., & Ferreira, E. Silagem ácida de resíduos de peixes para frangos de corte. *Acta Veterinaria Brasilica* (2016) 10(3), 284–289.

Chapter 6

Technical-Economic Analysis of the Use of Fish Waste Silage from Red Tilapia (*Oreochromis* Spp.) as Raw Material for Laying Hens' Feed

Yhoan S. Gaviria
Jairo A. Camaño
and José E. Zapata*, PhD
Department of Food and Pharmaceutical Sciences, Food Nutrition and Technology Group, Universidad de Antioquia, Medellín, Colombia

Abstract

One of the greatest challenges that agricultural systems currently face is the adequate treatment of solid waste because its generation rate increases every year at a rate that exceeds the capacity to properly dispose of it. One of the agricultural wastes with the highest environmental impact comes from fish subproducts, and using silage has been applied as a strategy for lipid and protein reassessment for them. In this study, Cost-benefit analysis was applied to assess the technical-economic feasibility of a production system consisting of a red tilapia fish farm and a poultry farm. The 3 farming production systems studied were: tilapia, tilapia and poultry fed with conventional raw materials, and tilapia and poultry fed with viscera silage and other raw materials. Parameters such as investment costs, productivity, and balance points were determined in addition to financial indicators such as net present value, internal return rate, cost-benefit ratio, and investment payback period. Results showed that using chemical silage

* Corresponding Author's E-mail: edgar.zapata@udea.edu.co.

In: Agro-Industrial Wastes
Editor: Peter Clements
ISBN: 979-8-89113-688-5

from red tilapia viscera in laying hens' feed is an economically feasible alternative in the constitution of a fish and poultry production system that generates higher profits than conventional production systems. Additionally, it also reduced the environmental impact generated by this waste from the fish industry.

Keywords: chemical silage, *Oreochromis* spp, Isa Brown, investment cost

Introduction

One of the greatest challenges for agricultural systems currently is the optimal treatment of solid waste, given that its generation rate increases every year at a pace that exceeds the capacity of competent authorities to properly control and manage it (You et al., 2016). Fish farming produces significant amounts of subproducts during its processing, making it one of the sectors that generates the highest environmental load (Rabassó and Hernández, 2015). This occurs because, according to the species, filleting yield is in the range of 30–50% (Arias et al., 2017). This is one of the industries with the highest production and constant growth lately worldwide, which makes it one of the most important industrial sectors (FAO, 2018).

The traditional methods established for treating fish industry waste focus on big producers, leaving small and medium fish production cooperatives with inefficient waste treatment, and they are inadequately spilled by river shores and buried in shallow dwell (Mota et al., 2019). Recently, fish silage has been implemented as an alternative method to use and process the waste from these fish and reduce the environmental impact they cause (Davies et al., 2020). The process consists of adding organic or inorganic acids or some acid-lactic bacteria to progressively reduce pH and activate endogenous enzymes in the waste, aiming to cause protein lysis, favoring the formation of peptides and amino acids of interest (Olsen et al., 2017). Additionally, it causes the reduction of pathogenic microbial flora (Olsen et al., 2017). Also, one of the main features of silage is its protein concentration, which, thanks to its reduction in size, has better digestibility and makes it a relevant alternative as protein raw material for various animal species feed (Davies et al., 2020).

Fish waste chemical silage has been studied from a technological, nutritional, and productive perspective, in addition to its implementation as an alternative protein source for various animal species feed (Banze et al.,

2017; Davies et al., 2020; Gaviria et al., 2020; Camaño et al., 2021). However, there is limited knowledge regarding the financial or economic feasibility assessment of the implementation of an industrial process for chemical silage from red tilapia viscera as part of a conventional fish farm system. Technical-economic assessment is essential, as it provides the economic profile of a productive project, allowing organizations or producers to make assertive decisions from a financial perspective (Živković et al., 2017). Additionally, different research studies use pilot approaches that enable obtaining the maximum amount of information at a lower cost about the economic profile of a productive model. This allows for large-scale estimations to support investment decisions (Farid et al., 2020).

Cost-benefit analysis (CBA) is one of the financial tools most used for the technical-economic assessment of production systems related to waste treatment (Di Trapani et al., 2014). This methodology enables a thorough assessment of a project's costs and benefits (program, intervention or policy measurement), aiming to determine whether the project is desirable for social wellbeing and, if it is, to what extent. This tool is intended to determine different economic patterns related to the purpose of the productive system, such as net present value (NPV), internal return rate (IRR), cost-benefit ratio (CBR), and investment payback period (IPP), which have already been implemented in the economic assessment of the use and exploitation of agro-industrial waste (You et al., 2016; Khootama et al., 2018; Weber et al., 2020). Therefore, the purpose of this study was to apply the cost-benefit analysis to assess the technical-economic feasibility of the system that integrates red tilapia (*Oreochromis* spp.) production, the manufacture of chemical silage from its viscera, and its use in the feed for Isain -Brown laying hens.

Methodology

This study was carried out based on the average production of small and medium-sized fish farming communities in Colombia, considering the income and costs of raw materials implemented along the whole production process. Accordingly, the methodology includes several subsections.

Study Systems

A number of three production systems were proposed based on results obtained in previous studies of the research group that developed this current study (Gaviria et al., 2020, Gaviria et al., 2021; Camaño et al., 2020; Suarez et al., 2018; Arias et al. 2017). The first system consisted of the industrial use of subproducts obtained in the evisceration stage of a red tilapia (*Oreochromis* spp.) production farm as a starting point for the variables in the technical-economic feasibility assessment of the production system. On the other hand, the traditional system for assessment comparison was considered. Table 1 describes the three systems evaluated. Additionally, Figure 1 shows the three process diagrams with their corresponding stages: Tilapia production (CFF), production of tilapia and laying hens fed with conventional raw materials (CFPF), and production of tilapia and laying hens fed with silage and other raw materials (SFPF).

Table 1. Description of the farm production systems evaluated

System	Description and characteristics
CFF	Control fish farm. This system consisted of a conventional fish farm producing red tilapia (*Oreochromis* spp.) without waste utilization.
CFPF	Control fish and poultry farm. This system was based on the production of tilapia and laying hens of the Isa breed fed with conventional raw materials.
SFPF	Fish-poultry farm with production of chemical silage. This system consisted of a fish farm linked to the feeding of laying hens of the Isa-Brown breed, using diets made from chemical silage.

Fish Production

This production system involved farming 32200 red tilapia (*Oreochromis* spp.) fingerlings for 7 months, which brought 28040 adult fish for commercialization, with a death rate of 13% (Gonzales and Quevedo, 2018, FAO 2009). For this process, six geomembrane tanks are required for the pre-fattening stage during the first 3 months with a density of 50 fish/m^3 and a tank diameter of 4.5 m. Next, 8 tanks will be needed for the fattening stage, with a density of 50 fish/m^3 and a tank diameter of 7.3 m for the 4 remaining months. The monthly production of fish is 4004 fish with an average weight of 0.40 kg (FAO 2009). The monthly number of viscera produced was estimated according to the percentage of the species; in the case of red tilapia, it is between 12% and 18% (Villamil et al., 2017), thus obtaining

256.4 kg (16%) (Suarez et al., 2018), which will be disposed of differently according to the farm production system.

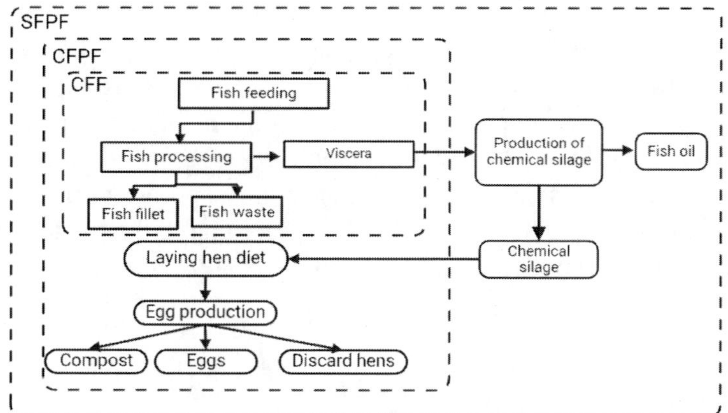

Figure 1. Diagram of the different production systems.

Silage Production

Viscera waste was used to manufacture chemical silage and collect fish oil. This process involves the stages of degreasing, grinding, silaging, and storing, as reported by Suarez et al., (2018) and Camaño et al., (2020). The SFPF system uses wet silage to make diets for poultry feed. Silage was implemented according to the viscera obtained, and the feed amount required for feeding laying hens (Gaviria et al., 2020).

Diet Production

This stage consists of the production of feed for laying hens. The process is found in the SFPF and CFPF systems, and involves the stages of mixing, pelletizing, and drying. The formulation of the diets used, and their nutritional composition, were based on previous studies carried out and published by the research group of the authors (Gaviria et al., 2020). Silage produced monthly (182 kg) was used to manufacture 262 kg of feed for 80 hens for said month, with an inclusion of 32% of wet silage in the diet (Gaviria et al., 2020).

Poultry Production

The parameters used for hen production were set according to the experiments carried out by Gaviria et al., (2020), which evaluated the inclusion of red tilapia viscera silage in laying hens' feed. This system took 80 18-month-old laying hens (Gaviria et al., 2020) that are fed twice a day and have a constant supply of water for a continuous period of one year, when they reach the end of their shelf life and are sold as discarded. The poultry production cycle lasts for 7 months, a period in which the hens lay between 14364 and 13608 eggs per cycle, according to the productivity rates reported by Gaviria et al., (2020) for SFPF and CFPF systems, respectively.

Costs

Costs were classified into 3 groups: investment costs (1), variable costs (2), and fixed costs (3). The estimation of the different costs associated with each of the production systems was based on the calculation of a 7-month production period. Investment costs are defined as the sum of materials, supplies, and equipment required to start a production system (Weber et al., 2020). Such information was provided by suppliers and budget quotes, in which the indirect workforce needed for physical, hydraulic, and electric infrastructure is included. Variable costs constitute production costs according to the volume produced and are calculated as the sum of animals, raw materials, direct workforce, industrial services, and indirect manufacture costs (IMC), which include provisions and supplies, among others (Arora et al., 2018). Transportation costs were considered non-significant in comparison with the variable costs involved, according to studies associated with waste treatment (Bernstad and Jansen, 2011). Fixed costs were determined as the sum of the costs of assets' depreciation using the straight-line method for a 10-year period (Weber et al., 2020; Mota et al., 2019); maintenance of equipment, which was considered 0.01% according to You et al., (2016) and Weber et al., (2020); rent, and cleaning and administrative expenses. The data presented were calculated based on the current prices of the market in Colombian pesos (COP) and were converted into American dollars (USD), with a conversion rate of 3999 COP per USD (Banco de la República de Colombia, 2023).

Revenues

The methodology proposed by Weber et al. (2020) was used to determine market average prices, which gave the income from the sales of the products generated in the different production systems as a result, using as a market baseline 1 kg of fish; the classification of eggs: A, B, or C; 1 kg of chicken manure; 1 kg of oil fish; and one 70-week discarded hen, respectively, for each case.

Balance Points Estimation

Balance point (BP) is that where total income (TI) and total costs (TC) of the production are equal (IT=CT). This means there is neither profit nor loss, and their benefit from them is zero. The balance point for each of the products described in the income section was determined using the weight contribution margin (WCM), based on the calculation of the contribution margin of each product (WCM_S) (Potkany and Krajcirova, 2015). Equation 1 shows the sales participation percentage of each product ($\%PP_s$) regarding the sales total, whereas the unit contribution margin (UCM) was calculated with equation 2. WCM_S was calculated with equation 3, while WCM was estimated by adding WCM_S. Finally, equations 4 and 5 were used to calculate the total balance point (BP_T) and BP in units for each product (BP_s), respectively, where TFC corresponds to total fixed costs.

$$\%PP_S = \frac{Sales_s}{Total} \tag{1}$$

$$UCM = P_{p.u} - CV_{p.u} \tag{2}$$

$$WCM_S = UCM.\%PP_S \tag{3}$$

$$BP_T = \frac{CFT}{WCM} \tag{4}$$

$$BP_s = BP_S.\%PP_s \tag{5}$$

Cost-Benefit Analysis

This analysis is implemented to identify risks and facilitate accurate financial decision-making in investment projects. It is based on the calculation of financial indicators, which are the result of the interaction of all the project components (De Clercq et al., 2017). These analyses are based on the factorial method, relying on the utility of projected annual sales (Farid et al., 2020). Equation 6 (You et al., 2016) was implemented to calculate the net present value (NPV), where LT is the number of periods in which, for study purposes, cash flows up to 5 years were estimated. Net cash inflow (Ct) was calculated after updating the revenues and expenses, considering an average annual inflation rate of 4.6% (Banco de la República de Colombia, 2020). Investment costs (Co) for each production system were used, and the minimum acceptable rate of return (MARR), also known as the discount rate, was estimated using equation 7, where "i" corresponds to the yield percentage required by the investors, which is a function of the risk, and for this case, an average risk of 10% was used. On the other hand, average annual inflation "f" shows values similar to the discount rates of 15%, as reported by Manioğlu and Yilmaz (2006) in the economic analysis of this study.

$$NPV = \sum_{t}^{LT} \frac{C_t}{(1+MARR)^t} - C_o \tag{6}$$

The Internal return rate (IRR) refers to the profit or loss percentage that an investment or project will have and corresponds to the discount rate when there is an NPV of zero. This was calculated using the algorithm provided by Matlab (License 914762, version 2018). The cost-benefit ratio (CBR) was determined using equation 8 (You et al., 2016), including the cost of expenses for year t (C_{et}). If NPV > 0, the production system model is accepted. If NPV = 0, decision-making makes no difference, which indicates that money invested in the project brings a profit at the same rate of interest of opportunity. Whereas, if NPV<0 makes the model is not feasible. If IRR>MARR, the model is accepted. If IRR<MARR, it is rejected, and if IRR=MARR, decision-making is irrelevant. For CBR, it is understood that, if CBR>0, there is a net profit, and for CBR<0, there are losses (You et al., 2016).

$$\mathrm{MARR} = \mathrm{i} + \mathrm{f} + (\mathrm{i.f}) \tag{7}$$

$$\mathrm{CBR} = \frac{\mathrm{NPV}}{\sum_{t}^{LT} \frac{C_{et}}{(1+\mathrm{MARR})^{t}} + C_{o}} \tag{8}$$

Finally, the investment payback period (IPP), as defined in equation 9, is defined as the number of years required for the deducted cash flows accumulated to be equal to the total costs and the initial investment (Di Trapani et al., 2014). IPP is a measure of how soon the investment can start to generate profit. Therefore, it is desirable to be as low as possible (Engelberth, 2020). To calculate it, the previous period in which the investment payback "A" was used; the initial investment value corresponds to "b," "c" is the cash flow for period "A," and "d" is the cash flow of the period when the investment is paid back.

$$\mathrm{IPP} = \mathrm{A} + \left[\frac{\mathrm{b-c}}{\mathrm{d}}\right] \tag{9}$$

Results and Discussion

Tables 2 and 3 show the investment costs associated with the initial supplies in each of the production systems. They provide details on the quantity, unit cost, and total cost of the subsystems mentioned. These values were implemented as the baseline to estimate total investment costs in each of the production systems assessed (Table 4). Regarding investment, the highest cost corresponds to the fish farming system, which is also the main activity. This is due to the conditioning and installation of tanks with geomembrane, which represent over 65% of the total cost of the fish farm. The additional processes in SFPF and CFPF production systems for manufacturing diets, laying hens' feed, and processing fish farm waste to obtain chemical silage generate a raise in the initial investment when compared with the CFF system, which is restricted to producing only fish meat.

Table 2. Investment costs for fish farming facility and diet processing plant

Fish farm			
Concept	Quantity	Cost/u	Total cost
G1 Tank	6	784.20	4705.18
G2 Tank	8	1064.27	8514.13
Hydraulic system	2	88.78	177.56
Water pumps	4	48.01	192.05
Filters	4	853.42	3413.68
pH meter	2	9.00	18.00
Oxygen sensor	2	22.26	44.51
Industrial balance	2	40.01	80.02
Baskets	12	2.25	27.01
Basket holder	2	47.51	95.02
Tables	6	47.51	285.07
Freezer	2	1008.25	2016.50
Fishing equipment	2	57.51	115.03
Pressure washer	2	75.02	150.04
Labor	1	374.09	374.09
SubTotal			20207.90
Chemical ensiling process			
Gas kettle	1	500.13	500.13
Gas pipette	1	12.50	12.50
Food Processor	2	15.75	31.51
Plastic containers	4	2.75	11.00
SubTotal			555.14
Diet processing plant			
Mixer	1	950.24	950.24
Pelletizer	1	1500.38	1500.38
Solar dryer	1	130.03	130.03
Traditional mill	1	45.01	45.01
SubTotal			2625.66

Table 3. Investment costs for fish farming facility and diet processing plant

Concept	Unit	Cost/u (quantity)	Total cost
Zinc Tile	unit	4.13 (10)	41.26
Tie-downs for roof tiles	package	3.98 (6)	23.86
Steel studs	unit	1 (24)	24.01
Angeo mesh	meters	0.45 (60)	27.01
Curtains	meters	0.63 (24)	15.00
Bulbs	unit	1 (8)	8.00
Others	--	--	60.57
Cage modules	unit	150.04 (1)	295.07
Scale	unit	3.75 (1)	7.50
Transportation	--	25.01	25.01
Compost	--	15	15
Labor	--	150.04	150.04
SubTotal		692.32	

Table 4. Total investment costs (USD) for each integrated farm system

Systems	Investment costs (USD)				
	Fish farming	Chemical silage	Diets	Poultry	Total investment
CFF	20207.90	0.00	0.00	0.00	20207.90
CFPF	20207.90	0.00	2625.66	692.32	23525.88
SFPF	20207.90	1110.28	2625.66	692.32	24636.16

Production Costs

The addition of fixed and variable costs represents production costs (PC), which are detailed in Table 5. The values found showed that the CFF production system has the lowest fixed and variable costs. This is because it has fewer productive stages, whereas SFPF and CFPF systems involve more operational activities and more resources and services for their proper operation as they incorporate the poultry process into their production systems. However, the CFF system requires an external service for the collection and disposal of organic waste (viscera) according to the provisions of Colombian regulation in Decree 838 of 2005 (Republic of Colombia, 2005). This cost is $ 0.77 per kg of waste collected, while the SFPF system uses this waste to obtain chemical silage and fish oil that can be commercialized. Additionally, silage can also be used as a raw material to manufacture poultry feed diets, changing the costs of disposal into an additional income. On the other hand, SFPF and CFPF production systems have differences in their variable costs, mainly in industrial services and the raw materials used for the different feeds. This is mostly because the conventional system does not use the fish industry subproducts and requires an external service for its operation, as the CFF system does, and additionally, it uses a higher percentage of traditional protein raw materials such as soy cake and fish flour, which increase production costs for laying hens' feed.

It is important to mention the high percentage that feed costs represent in all three production systems (Table 6), which have similar tendencies to those reported in different studies of fish farming systems (Bozoglu and Ceyhan, 2009; Hadelan et al., 2012; Di Trapani et al., 2014). It is observed that feed production costs in the SFPF production system have an approximate increase of 7.2% compared with the CFF system due to the raw materials needed to produce feed for laying hens. However, that cost is lower than the one from the CFPF system, at 2.3% and 9.5% higher than the CFF farm. Additionally, when services between CFPF and SFPF systems are compared, it is observed that the second shows a considerable reduction of 62.8% considering the first one, due to the savings in the cost of waste collection services associated with them, which were included in the diets of poultry species, thus reducing production costs and, consequently, generating higher profits.

Table 5. Fixed and variable costs for the different production systems per cycle

Fixed costs						
Cost concept	CFF		CFPF		SFPF	
	USD	%	USD	%	USD	%
Depreciation	2313.94	9.78	2637.03	9.93	2507.49	9.17
Maintenance	231.39	0.98	263.70	0.99	250.75	0.92
Lease	150.04	0.63	150.04	0.56	150.04	0.55
Administrative	25.64	0.11	29.06	0.11	29.06	0.11
Cleanliness	101.43	0.43	114.93	0.43	114.93	0.42
Subtotal	2822.45	11.92	3194.75	12.03	3052.27	11.16
Variable costs						
Feed-fish	7547.22	31.89	8197.55	33.14	8347.36	32.52
Poultry feed	0.00		546.49		546.49	
Inputs-CS	0.00		58.71		0.00	
Subtotal, Food-Inputs	7547.22		8802.75		8893.85	
Animals	1046.76	4.42	1366.84	5.15	1366.84	5.00
Labor	3746.50	15.83	3933.82	14.81	4098.51	14.98
Indirect manufacturing costs	52.51	0.22	112.60	0.42	110.85	0.41
Industrial Services	906.42	3.83	347.18	1.31	934.98	3.42
Subtotal	13299.41	56.19	14563.19	54.83	15405.03	56.32
Total production costs	23669.08		26560.70		27351.15	

On the other hand, the value of the diets of the production system using the chemical silage from red tilapia viscera (SFPF) has a reduction of 31.7% ($253.65) compared with the conventional production system (CFPF). This is mostly because fish protein has the highest cost of raw materials for feed manufacturing, which is reduced by including 32% of silage in the SFPF production system, as reported by Gaviria et al., (2020). Moreover, considering that infrastructure and facility building costs represent the greatest part of costs in these systems, and that the deduction rate and all other concepts of costs are less controllable, strategies to help reduce operation costs must be proposed to improve the economy of the systems evaluated, giving special attention to CBR values to use waste for the production of chemical silage for different animal species feed.

Table 6. Profit and loss statement for the different production systems by cycle

Concept	CFF	CFPF	SFPF
Revenues (+)	24423.30	26368.15	26029.91
Fish	24423.30	24423.30	24423.30
Oil	--	284.97	--
Eggs	--	1103.23	1049.96
Chicken manure	--	504.13	504.13
Discard hens	--	52.52	52.52
Fixed costs (-)	2822.45	3194.75	3052.27
Depreciation	2313.94	2637.03	2507.49
Maintenance	231.39	263.70	250.75
Lease	150.04	150.04	150.04
Administrative	25.64	29.06	29.06
Cleanliness	101.43	114.93	114.93
Variable costs (-)	13299.41	13957.99	14858.54
Feed-fish	7547.22	7547.22	7547.22
Poultry feed	0.00	546.49	800.14
Inputs-CS	0.00	103.84	0.00
Subtotal, Food-Inputs	1046.76	1366.84	1366.84
Animals	3746.50	3933.82	4098.51
Labor	52.51	112.60	110.85
Indirect manufacturing costs	906.42	347.18	934.98
Gross profit	8301.44	9215.40	8119.10
Taxes (-)	2490.43	2764.62	2435.73
Net profit	5811.01	6450.78	5683.37

Income Analysis

Table 6 shows the income statement from the 3 production systems studied in a 7-month production cycle. The result indicates that the SFPF farm has the highest income level, followed by the CFPF, and the CFF goes last. Likewise, net utilities from the SFPF production system show the highest value compared with the other systems. This is because processing tilapia viscera to obtain chemical silage and oil fish, as well as the poultry production stage, generates new income that is reflected on the global statement. On the other hand, although the income from fish sales in CFF, SFPF, and CFPF corresponds to 100%, 92.62%, and 93.83%, respectively, it is worth noting that the SFPF production system that uses waste increased its net profit by 13.5% compared with the conventional CFPF system, whereas a small fish farmer who wants to invest in a small poultry system using conventional raw materials would have a decrease of 2.19% in their net profits.

These results show a high economic potential to implement these systems in small and medium producers, matching with the statements made by Mota et al. (2019), who designed a system to extract oil from the Nile tilapia viscera in Brazil, bringing the conclusion that coupling a system that involves technological processes to use both the protein and the lipid part of fish viscera can generate additional income and reduce the environmental impact caused by these organic macrocomponents. However, these results are not enough to determine the economic feasibility of the evaluated systems. Therefore, it is necessary to make a cost-benefit analysis (CBA), as carried out in this study.

Balance Points Analysis

The units needed to be commercialized to cover the total cost of a production system so that there are no economic losses during a set period are known as the balance point. Table 7 shows the balance points values for each of the products obtained in the three production systems proposed. The production stage that makes the biggest contribution is fish sale, which is over 1000 kg of fish in a 7-month period on all 3 farms. In the study cases, this tendency is because fish farming is the main productive activity, on the one hand, because of the production volume, and on the other, because it is the starting point for the other productive stages. The SFPF systems decreases by 105 units because of the inclusion of more products being sold, such as eggs and compost material, which would allow for reaching the balance point in a shorter time and generate a higher net profit per cycle, as discussed above. These results indicate that the profitability of integrated systems depends on the size of the plant used. Alternatively, the system with traditional feed for poultry species (CFPF) requires higher amounts of products to achieve the balance point for economic sustainability, which is correlated with the statement in the section of production costs given that this system is the one that demands the highest maintenance cost.

Table 7. Breakeven points for the different production systems

	Products			
System	Kg Fish	Kg Oil	# Eggs	Kg Chicken manure
CFF	2368	--	--	--
CFPF	2158	330	10150	1282
SFPF	3176	--	25238	3366

Egg production in the CFPF control system has issues complying with the projections of the balance point because just 15120 eggs are produced during the 7 months of a period. This happens mainly because they obtained negative values in the unit contribution margin (UCM) for these products. This means that unit variable costs are higher than unit sale costs, causing economic losses in the poultry model of this system. Consequently, it can be inferred that a small fish Farmer could not maintain a fish and poultry farming system of this capacity by using conventional raw materials for poultry feed, but if they implemented the SFPF system, economically feasibility could be achieved.

Cost-Benefit Analysis

Table 8 shows the technical-economic analysis using the cost-benefit assessment (CBA), which is a tool to validate sustainability and manage decision-making in study systems, as it allows comparing the feasibility of the production systems in terms of profitability and other indicators. NPV represents the total of resources in favor of the production system at the end of the period set, which was 5 years for this current study. This variable showed a positive value in the three systems studied, with the CFPF as the system with the lowest value in this parameter. Conversely, IRR values exceeded MARR values for all cases, showing that all farming systems valued can be economically feasible. However, regarding the net profit and production costs of the CFPF system, it was possible to determine that feasibility is mainly achievable in fish production because poultry production in this system does not generate profit. Therefore, the model would not be properly feasible compared with the conventional CFF system and the SFPF system based on the use of viscera. The system with the highest economic return is the CFF control system, followed by the SFPF system, which uses viscera waste. This was also a tendency for the CBR parameter, and the system that uses the subproducts from the fish industry was the one that obtained the best net profits compared with the CFPF farm.

Table 8. Financial evaluations for different production systems

System	NPV (USD)	IRR	CBR	IPP
CFF	15863.19	43.96%	0.12	1.99
SFPF	15406.24	38.50%	0.10	2.06
CFPF	11752.91	34.09%	0.08	2.12

However, the CFF system is shown to have the best financial solvency, as it presents the highest benefit when compared with the other two production systems. The CFPF system shows lower productive performance, with only 8% of net profit, proving that the fish/poultry multifunctional system with conventional raw materials has the lowest performance in the assessment of this study. The results obtained for the investment payback period (IPP) had a similar behavior to parameters NPV, IRR, and CBR for the farm that included silage processes, which clearly evinces a shorter payback time when compared with control farms.

Values obtained in the SFPF system both in IRR and RP are comparable to those reported by Mota et al. (2019), with an IRR and RP of 45% and 1.9 years, respectively, in the system using Nile tilapia (*Oreochromis niloticus*) viscera for oil fish extraction for biodiesel production. It is similar for the use of sweet potato waste, with an IRR and an RP of 51% and 1.06 years, respectively (Weber et al., 2020). However, some studies on the use of waste with lower results than the ones obtained in this study have been considered profitable and economically feasible, as in the case of the lipase enzyme of Aspergillus niger using agro-industrial mango waste, with an IRR of 35% (Khootama et al., 2018), and a biorefinery of mango waste processing, with an IRR of 34% (Arora et al., 2018). All these cases indicate that using waste generates environmental benefits and can be an economically feasible process. Nevertheless, there are cases in which waste use does not generate economic feasibility, as in the case of methane production from biological waste, whose NPV was negative (De Clercq et al., 2017) and a biorefinery of algae biomass to produce biofuel and energy, which was also unfeasible (Abdul and Lim, 2019). These negative results indicate that economic feasibility in waste use processes is not guaranteed and requires further assessment.

The results of the SFPF system show that economic benefit can be obtained with the production of DCS through the implementation of a poultry system fed with CS as a protein raw material to substitute conventional supplies. Currently, that information on the assessment of production systems that involve fish silage is limited, which makes the results of this study as a novelty contribution in the field of study. This is convenient as fish silage has been undergone several research studies (Davies et al., 2020), being characterized as an element with an important potential with high technical feasibility to substitute conventional protein raw materials that have a higher cost (Madage et al., 2015; Davies et al., 2020).

Finally, it has been established that there are difficulties in making comparisons with fish systems in other countries due to the particular aspects of fish farming in each country, which have significant changes (Mota et al., 2019). However, although the economic analysis in this study was based on references to costs and prices from the Colombian market, it is important to notice that using fish viscera through silage technology to implement them in conventional fish farming systems could be of great economic and environmental benefit for any producer in other countries.

Conclusion

Using chemical silage in red tilapia viscera for laying hens' feed in the combination of a fish and poultry production system is an economically feasible alternative. It will generate a higher profit for the conventional production systems as the investment payback period is reduced. On the other hand, the process to obtain chemical silage from red tilapia viscera has a lower production cost than for conventional proteins for poultry feed, which is a factor that reduces environmental impact and favors the economy of the production system.

Disclaimer

None

References

Abdul NN, Lim JS. Evaluation of processing route alternatives for accessing the integration of algae-based biorefinery with palm oil mill, *J. Clean. Prod* (2019) 212(2019): 1282–1299.

Arora A, Banerjee J. Process design and techno-economic analysis of an integrated mango processing waste biorefinery, *Ind. Crops. Prod* (2018) 116(February): 24–34.

Banze JF, Da Silva MFO, Enke, D.B.S., Fracalossi, D.M. Acid silage of tuna viscera: Production, composition, quality and digestibility, *Bol. Inst. Pesca* (2017) 43: 24–34.

Bernstad A, Jansen J. A life cycle approach to the management of household food waste - A Swedish full-scale case study, *Waste Manage* (2011) 31(8): 1879–1896.

Camaño JA, Rivera AM, Zapata JE. Efecto del espesor de película y de la ubicación de la muestra en un secador solar directo, sobre la cinética de secado de ensilado de vísceras de tilapia roja (*Oreochromis* sp), *Inf. Tecnol (2020)*31(1): 53–66.

Camaño JA, Rivera AM, Zapata Montoya JE. Sorption isotherms and thermodynamic properties of the dry silage of red tilapia viscera (*Oreochromis* spp.) obtained in a direct solar dryer, *Heliyon* (2021) 7(4).

Davies SJ, Guroy D. Evaluation of co-fermented apple-pomace, molasses and formic acid generated sardine based fish silages as fishmeal substitutes in diets for juvenile european sea bass (Dicentrachus labrax) production, *Aquaculture* (2020) 521(September 2019).

De Clercq D, Wen Z, Fei F. Economic performance evaluation of bio-waste treatment technology at the facility level, *Resour. Conserv. Recycl* (2017) 116: 178–184.

Di Trapani AM, Sgroi F, Testa R. Economic comparison between offshore and inshore aquaculture production systems of European sea bass in Italy, *Aquaculture* (2014) 434: 334–339.

Engelberth A. Evaluating Economic Potential of Food Waste Valorization: Onward to a Diverse Feedstock Biorefinery, *Curr. Opin. Green Sustain. Chem* (2020): 26.

FAO. The State of world fisheries and aquaculture, Meeting the Sustainable Development Goals (2018).

FAO. *Oreochromis niloticus in Cultured aquatic species fact sheets.* Text by Rakocy, J. E. Edited and compiled by Valerio Crespi and Michael New (2009). CD-ROM (multilingual).

Farid MAA, Roslan AM. Net energy and techno-economic assessment of biodiesel production from waste cooking oil using a semi-industrial plant: A Malaysia perspective, *Sustain. Energy Technol. Assess (2020)* 39(March).

Gaviria YS, Londoño FL, Zapata JE. Effects of chemical silage of red tilapia viscera (*Oreochromis* spp.) as a source of protein on the productive and hematological parameters in isa-brown laying hens (Gallus gallus domesticus), *Heliyon* (2020) 6(12).

Gaviria YS, Figueroa OA, Zapata JE. Efecto de la inclusión de ensilado químico de vísceras de tilapia roja (*Oreochromis* spp.) en dietas para pollos de engorde sobre los parámetros productivos y sanguíneos, *Inf. Tecnol* (2021) 32(3).

Gonzales C, Quevedo E. *Cultivo de las tilapias roja (Oreochromis spp.) y plateada (Oreochromis nilotus)* (p. 17) (2018).

Hadelan L, Par V, Njavro M, Lovrinov M. Real option approach to economic analysis of European sea bass (Dicetrarchus labrax) farming in Croatia, *Agric. Conspec. Sci* (2012) 77(3): 161–165.

Khootama A, Putri DN, Hermansyah H. Techno-economic analysis of lipase enzyme production from Aspergillus Niger using agro-industrial waste by solid state fermentation, *Energy Procedia* (2018)153: 143–148.

Madage SSK, Medis WUD, Sultanbawa Y. Fish Silage as Replacement of Fishmeal in Red Tilapia Feeds, *J. Appl. Aquac* (2015) 27(2): 95–106.

Manioğlu G, Yilmaz Z. Economic evaluation of the building envelope and operation period of heating system in terms of thermal comfort, *Energy Build (2006)* 38(3): 266–272.

Mota FAS, Costa JT, Barreto GA. The Nile tilapia viscera oil extraction for biodiesel production in Brazil: An economic analysis, *Renew. Sust. Energ. Rev (2019)* 108(March): 1–10.

Olsen RL, Toppe J, Karunasagar I. Fish silage hydrolysates not only a feed nutrient, but also a useful feed aadditive, Trends. *Food. Sci. Tech* (2017) 66: 93–97.

Potkany M, Krajcirova L, (2015) Quantification of the volume of products to achieve the break-even point and desired profit in non-homogeneous production, *Procedia Econ. Financ,* 26(15), 194–201.

Rabassó M, Hernández JM. Bioeconomic analysis of the environmental impact of a marine fish farm, *J. Environ. Manage (2015)* 158: 24–35.

República de Colombia, (2005) *Decreto 838 de 2005, Por el cual se modifica el Decreto 1713 de 2002 sobre disposición final de residuos sólidos y se dictan otras disposiciones.*

Suarez LM, Montes JR, Zapata JE. Optimización del contenido de ácidos en ensilados de vísceras de Tilapia roja (*Oreochromis* spp.) con análisis del ciclo de Vida de los alimentos derivados, *Inf. Tecnol* (2018) 29(6): 83–94.

Villamil O, Váquiro H, Solanilla JF. Fish viscera protein hydrolysates: Production, potential applications and functional and bioactive properties, *Food Chem* (2017) 224: 160–17.

Weber CT, Trierweiler LF, Trierweiler JO. Food waste biorefinery advocating circular economy: Bioethanol and distilled beverage from sweet potato, *J. Clean. Prod* (2020) 268.

You S, Wang W. Comparison of the co-gasification of sewage sludge and food wastes and cost-benefit analysis of gasification- and incineration-based waste treatment schemes, *Bioresour. Technol* (2016) 218: 595–605.

Živković SB, Veljković MV. Technological, technical, economic, environmental, social, human health risk, toxicological and policy considerations of biodiesel production and use, *Renew. Sust. Energ. Rev (2017)* 79(February): 222–247.

Chapter 7

Implementation of the Ecological Footprint Methodology as an Indicator of Sustainability in the Use of Fish Waste for the Feeding of Laying Hens

Yhoan S. Gaviria Gaviria[1], PhD Candidate
Luis F. Londoño Franco[2], PhD
and Jose E. Zapata Montoya[1,*], PhD
[1]Nutrition and Food Technology Laboratory, Universidad de Antioquia, Medellin, Antioquia, Colombia
[2]Aquaculture Research Group, Politecnico Jaime Isaza Cadavid, Medellín, Antioquia, Colombia

Abstract

In recent years, global fish production has risen due to population growth and increased consumer interest in this product, leading to a corresponding increase in generated waste. This is particularly evident in the case of fish viscera, a direct contributor to negative environmental impacts. The ecological footprint methodology stands out as a widely utilized sustainability indicator for assessing the environmental impact of processes. This method quantifies the impact of human activities on the environment, encompassing factors such as water usage, forest products, infrastructure, and carbon footprint, thereby providing comprehensive, comparable, and reliable results. This study focused on determining the environmental impact of utilizing red

* Corresponding Author's E-mail: edgar.zapata@udea.edu.co.

In: Agro-Industrial Wastes
Editor: Peter Clements
ISBN: 979-8-89113-688-5

tilapia (*Oreochromis* Spp.) viscera for producing chemical silage and incorporating it into the diet of laying hens, using the ecological footprint methodology as a sustainability indicator. The productive system analyzed was a red tilapia (*Oreochromis* ssp.) farm situated in the municipality of San Jerónimo, Antioquia, Colombia. Productive variables of the laying hens, including laying percentage, egg weight, and feed conversion rate, were evaluated. The production of chemical silage in this process resulted in a monthly reduction of 1.493 kg of CO_2 compared to the emissions from discharging fresh viscera into shallow dumps. Notably, the key categories exerting the most significant impact on the production system were the utilization of natural resources and the disposal of wastewater. Conversely, the productive variables of Isa Brown breed laying hens were not significantly affected by the inclusion of chemical silage, maintaining the laying percentage and improving feed conversion. In conclusion, utilizing fish by-products, such as red tilapia (*Oreochromis* Spp.) viscera, for laying hen feed production leads to a reduced environmental footprint compared to traditional waste disposal methods. The chemical silage from red tilapia viscera can serve as an alternative protein source in laying hen feed manufacturing without altering production parameters while enhancing feed conversion.

Keywords: ecological footprint, fish waste, layer hen, chemical silage, wastewater

Introduction

Human agricultural practices are closely linked to the depletion of natural resources, production of waste, and emissions of greenhouse gases (Ahmed & Wang, 2019). Aquaculture, particularly fish farming, has seen remarkable growth over recent years, both in productivity and economic value, with a global output of 171 million tons in 2016. In Colombia, aquaculture production was projected to grow by 9% in 2018, with red tilapia (*Oreochromis* spp.) accounting for 62% of this increase (FAO, 2018). However, this industry also produces a significant amount of waste, approximately 65% of its total production, much of which is improperly disposed of, leading to environmental harm (Martínez-Alvarez et al., 2015). A notable portion of this waste, especially fish viscera, is rich in macronutrients and could be repurposed as a valuable protein source for animal feed (Gaviria G et al., 2020).

The cost of feed in poultry farming, which has emerged as the fastest expanding sector in livestock farming due to high demand, represents a major challenge. This is because dietary expenses, especially those for energy and protein, account for 95% of the total feeding costs (Farrell, 2013). The primary sources of protein in these diets are fish and soy flour, with corn flour serving as the main energy source (Ravindran, 2013). However, the high demand for these ingredients, coupled with their limited availability in countries like Colombia, necessitates expensive imports and raises production costs, prompting the search for alternative protein sources. One such alternative is the silage made from fish viscera, which is rich in proteins and lipids, remains stable for several months at room temperature, and meets the nutritional needs of various animal specics (Gaviria G et al., 2020). Despite its benefits, the process of producing chemical silage involves the use of organic acids, which may have adverse environmental impacts. This raises the need to evaluate the environmental consequences of using chemical silage in animal feed. The ecological footprint method, a widely recognized sustainability indicator, offers a comprehensive means to assess the environmental impacts of human activities (Gwehenberger & Narodoslawsky, 2007). It measures the effect on water, forests, infrastructure, and carbon emissions, providing a holistic, comparable, and reliable analysis (Ahmed & Wang, 2019). This study aims to assess the environmental impact of using chemical silage made from red tilapia viscera in the diet of laying hens, employing the ecological footprint as a measure of sustainability.

Materials and Methods

Study Location

The production system was based at a red tilapia (*Oreochromis* spp.) farm situated in San Jeronimo, Antioquia, Colombia, at coordinates 6°26′30″N, 75°43′40″W. This location experiences temperatures between 18 and 25 degrees Celsius, maintains an average humidity of 80%, and receives annual rainfall ranging from 1,000 to 4,000 mm. The proposed method involved processing the viscera obtained from tilapia harvests through chemical silage, with the intention of utilizing this processed material as feed for laying hens (Gallus gallus domesticus).

Production System

Figure 1 outlines the boundaries of the production system, detailing both the outputs and waste generated at each stage.

Sub-system 1 (Subs1) encompasses the initial phases of red tilapia (*Oreochromis* spp.) cultivation, starting with 2,044 fingerlings. These fish are raised over a seven-month period until they reach an average weight of 400g, during which they are fed three types of concentrated feed tailored to their specific stages of growth.

Sub-system 2 (Subs2) involves the processing of the fish. In this phase, 1,778 fish, each weighing 400 grams, are processed. This number reflects a mortality rate of approximately 13% during the breeding and fattening stage. During processing, the viscera, which constitute 16% of the total weight of the fish, are identified as waste.

Table 1. Formulation of layer hen diets (%)

Raw Material	Percentage
Soybean meal	8.10
Fish meal	6.85
Chemical silage	21.91
Corn meal	35.34
Rice flour	13.06
Fish oil	1.00
Calcium carbonate	8.09
Dicalcium phosphate	4.04
Vit. min. Supplement*	0.40
Methionine	0.30
Lysine	0.30
Tryptophan	0.30
Threonine	0.30

* Composition per 250 g of product: vit. A - 1.400.000 IU; vit. B1 - 500 mg; vit. B12 - 300 mg; vit. B2 = 500 mg; vit. B6 – 1.6 g; vit. D3 - 2.500.000 IU; vit. E - 6.000 IU; vit. K3 = 1.000 mg; biotin - 30 mg; niacin -12 g; folic acid - 1 g; cobalt - 50 mg; copper - 3.000 mg; iron - 25 g; iodine - 500 mg; manganese - 32.5 g; selenium - 100.50 mg; zinc - 22.49 g.

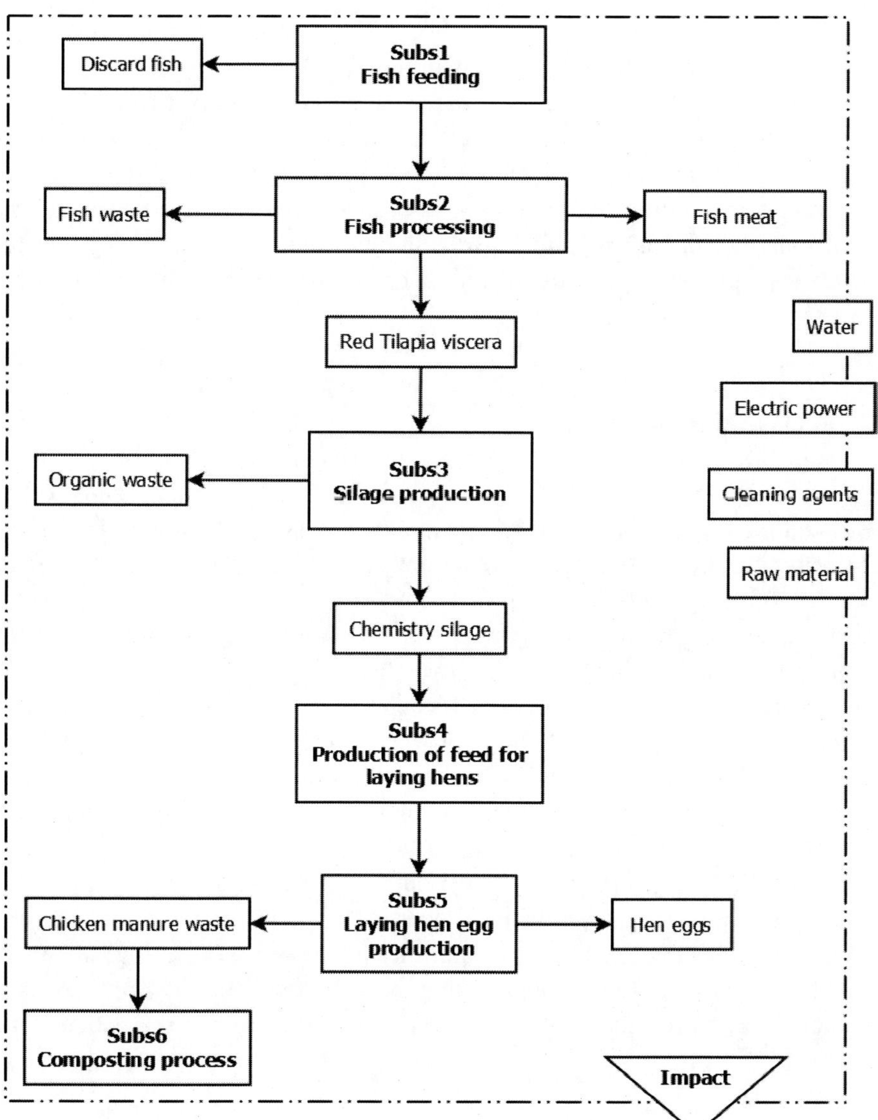

Figure 1. Limits of the production system.

Sub-system 3 (Subs3) is dedicated to the creation of chemical silage (CS) from the viscera waste, following the procedure outlined by Suarez et al. This involves several steps: degreasing, shredding, fermenting into silage, and then storing the final product. All of the chemical silage produced in this process is allocated for the diet of laying hens.

Sub-system 4 (Subs4) involves the preparation of feed for the laying hens, which is executed in three main stages: mixing the ingredients, pelletizing the mixture, and then drying the pellets to complete the feed. The specific formulation of the poultry feed (Table 1) is determined based on the nutritional needs outlined in the Brazilian nutrition tables, ensuring that the diets are optimized for the health and productivity of the laying hens (Rostagno et al., 2011).

Animals and Management

Sub-system 5 (Subs5) deals with the egg production process, utilizing 36 Isa-Brown breed laying hens that are 16 weeks old. These hens were fed over a 7-week period and divided into two groups: a control group that received commercial feed and a silage group. The silage group was further divided into smaller subgroups, each containing 3 hens. Feeding occurred from the 16th to the 32nd week of the hens' lives. For the purposes of evaluating the environmental impact, the analysis was conducted over a one-month period, whereas the productivity of the hens was assessed over the entire 7-week duration. The hens received feed twice daily, in the morning and afternoon, with each bird being allocated a 107-gram portion of feed, alongside constant access to water.

Sub-system 6 (Subs6) is focused on the composting of waste produced from the hens during the feeding process. This involves the collection and composting of the excretions from the poultry, transforming it into a valuable organic fertilizer, thus minimizing waste and contributing to a more sustainable cycle of nutrient management within the production system.

Assumptions and Limitations

In conducting this study, several key assumptions were made to streamline the analysis and focus on the core aspects of the environmental impact. These assumptions include:

1. The process of making Chemical Silage (CS) is integrated within the fish farming operations on the same premises. This strategic

decision is aimed at mitigating the environmental footprint associated with transporting fish viscera to a separate location for processing.

2. The environmental impact assessment excludes the effect of the infrastructure, such as buildings, involved in the production process. This exclusion is based on the rationale that the buildings have a prolonged lifespan, which dilutes their relative impact over time, rendering it negligible for the purpose of this analysis.
3. Similarly, the environmental contribution of industrial equipment used in the production process is not considered in the environmental impact calculations. Previous studies and analyses have shown that the long-term utilization of such equipment diminishes its relative impact, suggesting that including it in the assessment would not materially alter the findings (Perez-Martinez et al., 2018).
4. For Sub-system 4 (Subs4), where diets for laying hens are prepared, it is assumed that the only emissions released into the environment during the drying process are water vapor molecules. This assumption is predicated on the understanding that the operational temperatures during this stage are not high enough to cause the evaporation and subsequent release of other potentially harmful compounds.

These assumptions are intended to narrow the focus of the environmental impact analysis to the most significant factors, thereby providing a clearer understanding of the primary contributors to the ecological footprint of this integrated fish farming and egg production system.

Calculation of Environmental Impact

For the environmental impact analysis of the production system depicted in Figure 1, the ecological footprint methodology was employed. This approach involves using detailed calculation models to estimate the ecological impact at each stage of the system (Gwehenberger & Narodoslawsky, 2007). The advantage of this methodology lies in its precision and its ability to overcome the limitations encountered when using international databases that may lack specific regional adjustments (Mamouni Limnios et al., 2009). In the context of this analysis, the breeding of 2,044 red tilapia fish (*Oreochromis* spp.) was considered the base unit. The environmental impact of inputs and outputs from each stage of the production system was normalized to the same unit (Ha/ton), reflecting the area of forest land required in a particular region to sequester the CO_2 emitted at each stage of the system (Gwehenberger & Narodoslawsky, 2007).

The calculation of the environmental impact (IAco) attributable to organic compounds (co) generated or used in each stage of the production system was based on Equation 1, in line with the guidelines set forth by the Intergovernmental Panel on Climate Change (IPCC). This equation factors in the quantity of organic waste (CRO); the correction factor for methane gas (CH4), FCM, which varies depending on the solid waste management processes of the specific sector (IPCC, 2006); the fraction of the waste's degradable organic carbon (COD) that can undergo biochemical decomposition; the fraction of degradable organic carbon (CODF) that is not assimilated or degrades very slowly; the fractions of CH_4 in landfill gas (F) and the fraction of CH4 that is recovered (R), respectively (IPCC, 2006). The oxidation factor for methane (OX), the global warming potential of methane (PCG) for a 100-year period (IPCC, 2006), and the CO_2 capture factor for the region under study (FF) established in the environmental analysis (IDEAM, 2011), are also included in this equation.

$$IA=\sum_{co=1}^{n} CRO\left(\frac{16*(FCM*COD*COD_F*(F\text{-}R)*(1\text{-}OX)*PCG)}{12*FF}\right) \tag{1}$$

To calculate the environmental impact stemming from the electric energy input during various stages, Equation 2 was utilized. In this equation, EEG represents the electric energy expenditure in each stage (e) of the process, while EEE denotes the CO_2 emission factor associated with one kilowatt-hour (kw-h) of energy (IDEAM, 2011).

$$IA=\sum_{e=1}^{n} EE_G*\frac{EEE}{FF} \tag{2}$$

In sub-system 3, during chemical silage making stage, a degreasing process takes place by propane gas heating. In order to determine its impact, the calculation is performed using Equation 3, where VRC_3H_8 is the gas volume required for the process; EBp is the energy imbued in the process of obtaining propane gas; the distance covered in transportation, fuel mileage and propane gas loads are marked with DR, Rc and C, respectively. Finally, CO_2 emissions coming from fuel were marked as EC. Additionally, there is generation of CO_2 in that sub-system because of propane gas combustion process, for which its impact was calculated by Equation 4, where FCC3H8 corresponds to the conversion factor for propane gas, and FEC_3H_8 is the CO_2 emission factor (IPCC, 2006).

$$IA=VR_{C_3H_8}\left(\frac{EB_P{*}EEE+\left(\frac{D_R}{R_C{*}C}\right){*}EC}{FF}\right) \tag{3}$$

$$IA=\frac{VR_{C_3H_8}{*}FC_{C_3H_8}{*}FE_{C_3H_8}}{FF} \tag{4}$$

Including an input flow corresponding to cleaning water for equipment and spaces for processing is necessary in the different sub-systems. For this, the impact caused by water flow was calculated with Equation 5, where AL corresponds to the liters of water required in the process (a); and EBA is the energy imbued for water supply. However, using this brings an environmental impact associated to its final arrangement, which is calculated using Equation 6, where ARL corresponds to wastewater obtained in the cleaning process; $CODBO_5$ is the biochemical oxygen demand from the degradable fraction of wastewater; and $CMPCH_4$ indicates the maximum amount of methane production that such fraction has (IPCC, 2006).

$$IA=\sum_{a=1}^{n} A_L\left(\frac{EB_A{*}EEE}{FF}\right) \tag{5}$$

$$IA=\sum_{a=1}^{n} AR_L{*}\left(\frac{CO_{DBO5}{*}CMP_{CH_4}{*}FCM_{CH_4}{*}PCG}{FF}\right) \tag{6}$$

For the environmental impacts associated to production ingredients and other input flows in the different sub-systems, Equation 7 is used, where CR_I is the "i" product quantity required; EBP_I corresponds to the energy imbued for its manufacture; and C_I the maximum load in product transportation (IPCC, 2006).

$$IA=CR_I\left(\frac{EB_{PI}{*}EEE+\left(\frac{D_R}{R_C{*}C_I}\right){*}EC}{FF}\right) \tag{7}$$

Productive Variables

The egg laying percentage was evaluated by dividing the number of eggs laid each day by the total number of hens. This calculation provides a measure of the productivity of the hens in terms of egg production. Eggs were collected, counted, and weighed daily using a precision analytical scale (TxB220-1L from Shimadzu, Japan) with a precision of 1 gram. This process was carried out consistently over a period of 7 weeks to determine both the laying percentage and the average weight of the eggs. The feed conversion ratio (FCR) was assessed on a weekly basis by measuring the amount of feed consumed in kilograms and dividing it by the number of egg dozens produced. This ratio provides insight into the efficiency with which the hens convert feed into egg production, serving as a key indicator of the economic and resource efficiency of the egg-laying operation (Akinola & Iyomo, 2018).

Statistical Analysis

The data collected was evaluated at a confidence level of 95% by means of the hypothesis test to determine the difference in means using Fisher's LSD (Least significant difference) test with statgraphics centurion XVI software.

Results and Discussion

Productive Variables

The productive parameters of laying hens, which were fed with a partial protein substitution using chemical silage of red tilapia viscera, are summarized in Table 2. The egg laying percentage ranged between 82% (week 0-1) and 100% (week 5-6), which falls within the typical range for this particular line of laying hens (Silva et al., 2017). This consistency suggests that the inclusion of degreased chemical silage in the hens' diet did not have any detrimental effects on egg laying percentage. This outcome indicates that the nutritional composition of the diets was appropriately adjusted to meet the hens' requirements. Similar findings were reported by Kjos et al., (2001)., albeit with a 5% silage inclusion rate in their study.

Additionally, it was observed that egg weight increased progressively throughout the duration of the study, which aligns with the findings of Padhi et al., (2013) who noted that hen eggs tend to increase in weight up to the age of 52 weeks. Comparable results were reported by Batalha et al., (2018)who investigated the effects of water substitution in the original feed formulation, indicating that higher concentrations of silage did not significantly affect egg weight over time. Conversely, it is suggested that such substitutions may even enhance egg weight, particularly due to the quality of dietary lipids. Vitellogenesis and egg laying are known to be highly dependent on lipid and energy levels in the ovary (Coorey et al., 2015; Zhou et al., 2008).

Table 2. Productive parameters of laying hens

Weeks	Laying percentage (Control)	Laying percentage (Silage)	Egg weight (g) (Control)	Egg weight (g) (Silage)	FCR (kg feed/dozen eggs) (Control)	FCR (kg feed/dozen eggs) (Silage)
0-1	85.71 + 3.52	82.14 + 4.30	49.80 ± 1.47	49.67 + 3.51	2.28 ± 1.30	1.56 + 1.06
1-2	84.72 + 4-.54	85.71 + 5.81	55.23 ± 2.49	51.72 + 2.24	2.04 ± 1.34	1.50 + 2.30
2-3	85.22 + 2.13	82.14 + 4.53	58.14 ± 2.48	51.79 + 3.68	1.93 ± 1.54	1.56 + 1.35
3-4	80.29 + 1.92	85.71 + 2.76	58.28 ± 1.56	52.34 + 5.14	2.08 ± 1.42	1.50 + 1.94
4-5	85.71 + 5.42	85.71 + 4.69	57.90 ± 1.45	56.51 + 2.61	2.05 ± 1.29	1.50 + 2.26
5-6	86.20 + 3.17	100.00 + 3.26	60.14 ± 2.61	69.52 + 1.74	2.02 ± 1.67	1.28 + 2.08
6-7	86.22 + 3.68	89.29 + 3.18	58.85 ± 3.64	72.40 + 3.15	1.56 + 1.22	1.44 + 1.38

FCR: Feed conversion ratio.

Furthermore, the feed conversion ratio (FCR) ranged between 1.28 and 1.56 kg of feed per egg dozen. These values demonstrate a more efficient feed conversion compared to the results reported by Batalha et al., (2018), where FCR ranged from 2.01 to 1.56 kg per egg dozen. This suggests that the inclusion of chemical silage positively influences feed conversion efficiency in terms of egg production.

Quantification of Environmental Impact of Chemical Silage Use in Bird Feeding

The results obtained from the use of silage in feeding laying hens warrant the quantification of the environmental impact caused by this process. To facilitate this analysis, Table 3 presents the quantification of the environmental impact as well as the quantity of product or waste obtained in each sub-system of the process (see Figure 1). In the fish breeding stage (Subs 1), an average reduction of 13% in the initial population (2044 fish) was observed due to mortality. This mortality rate falls within the typical range for a productive cycle of this nature, which usually ranges from 10 to 20% (Alicorp, 2010). However, despite this being a typical range, this sub-system exhibits the highest environmental impact within the entire production system, totaling 0.574 hectares per month. This is primarily attributed to the significant quantity of feed consumed by the fish during their breeding phase, as well as the growth of certain raw materials used to meet this requirement, both of which are associated with high greenhouse gas emissions (Henriksson et al., 2015).

Table 3. Environmental impact of the production system

Subsystem	Quantity	Environmental Impact (Ha/month)
Subs 1	1778.28	0.574
Subs 2	561.94	0.468
Subs 3	80.73	-0.144
Subs 4	159.79	0.083
Subs 5	1350.00	0.179
Subs 7	200.00	-0.108
Total		1.054

During the fish growth stage (Subs 2), scales (5%) and viscera (16%) are removed, resulting in an average weight yield of 79%. This stage demonstrates a notably higher environmental impact compared to subsequent stages. This is primarily due to the substantial volume of water required during evisceration, leading to wastewater with a high organic load. In many cases, this wastewater is discharged into the environment, further exacerbating its environmental impact (Quiroz Fernández et al., 2018).

Sub-system 3 involves the process of obtaining chemical silage, which yields 71% after the degreasing stage, resulting in a monthly production of 80.73 kg of chemical silage (CS). This product serves as a protein substitute for conventional raw materials like soybean and fish flour in the manufacture of laying hens' feed. The environmental impact of this stage is calculated at -0.144 hectares per month. The negative value indicates that the exploitation of the by-product (viscera) results in a net reduction of adverse environmental impact equivalent to that magnitude. Similar findings were reported by Malakahmad (Malakahmad et al., 2017), who determined the environmental impact of various organic waste disposal processes, highlighting that CO_2 emissions avoided in anaerobic digestion processes outweigh the total CO_2 emissions, resulting in a negative value and indicating a favorable environmental outcome (Malakahmad et al., 2017).

Sub-system 4 involves the feed manufacturing process for laying hens, which exhibits relatively lower environmental impacts compared to other stages of the production system. The main impact categories in this stage include obtaining raw materials and the use of electric energy in concentrated feed pelletizing. To mitigate environmental loads, the drying process utilizes solar collectors, as reported by Camaño et al., (2020), aiming to reduce the environmental footprint of feed production. This shift towards renewable energy use in food drying processes aligns with the trend observed in the environmental performance assessment of several poultry farms and processing plants. This assessment has shown that feed production contributes significantly to environmental pressures due to high energy and heat consumption, as well as the use of raw materials like soybean, which require mineral fertilizers (Skunca et al., 2018).

The feed conversion ratio (FCR) holds significance in egg production from an environmental standpoint, as poultry digestion generates significant greenhouse gases such as methane. This ratio represents the amount of gas produced per feed unit consumed for a meat or egg unit produced (Ibidhi et al., 2017). Sub-system 5 presents the environmental impact values associated with laying hens' feeding, primarily influenced by feed consumption. This consumption directly affects the amount and composition of manure produced, thereby influencing emissions from the poultry industry's production system. Similar findings were reported by Perez López et al., (2019) who concluded that changes in food consumption and composition affect environmental impacts during growth production and food processing, as well as subsequent emissions from poultry manure during housing, storage, and field application.

In sub-system 6, poultry manure generated showed a negative value, indicating that it contributed positively to the environment rather than being a burden. This is because the manure was utilized as fertilizer, offsetting the need for synthetic fertilizers or nitrogen-sequestering rotational crops. The different physical and chemical characteristics of poultry excretions make them suitable for use either as fertilizer or animal feed (Perez López et al., 2019). However, when not properly exploited, they can emit gases such as ammonia (NH3), nitrous oxide (N2O), and methane (CH_4) during the productive, storage, and land application stages (Leinonen et al., 2012a).

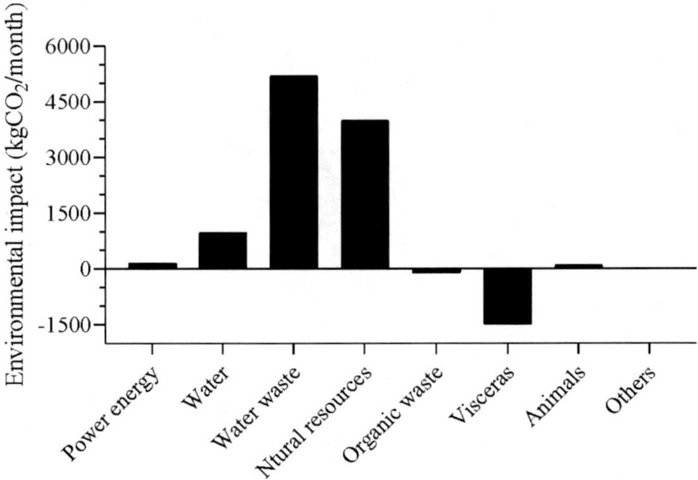

Figure 2. Environmental impact categories.

The main impact categories of sub-system inputs and outputs are depicted in Figure 2, with notable sources of greenhouse gas emissions represented in monthly kilograms of CO_2. The adoption of alternative energies, such as solar drying in the feed manufacturing process for poultry species, led to a significant reduction in CO_2 emissions in this category. Solar energy is a free, renewable, and abundant source, making it one of the most promising alternatives for reducing environmental impacts in drying processes. In contrast, the use of electric energy is typically associated with higher environmental impact indicators (Tiwari, 2016). Wastewater disposal emerges as the category with the highest environmental impact, primarily due to its organic load. The presence of organic matter makes wastewater disposal challenging as it requires large amounts of oxygen for degradation, diminishing its capacity to assimilate contaminant loads and naturally restore its quality (Quiroz Fernández et al., 2018). Additionally, the utilization of natural resources for raw material growth and supplies needed for both fish and poultry feed production represents another significant source of environmental impact in the production system. These findings align with those reported by Leinonen et al., (2012b) which highlighted the relatively high impact of concentrated feed production in the poultry industry, accounting for over 70% of the total global warming potential.

The utilization of organic waste from both the fish (viscera) and poultry (excretions) industries not only aids in reducing environmental impact directly but also generates added value to these by-products, thus positively impacting the economy of the process. While exploiting by-products from these industries leads to a reduction in environmental impact, it's important to acknowledge that in the processing of such residuals, new environmental impact sources may emerge. Therefore, complete elimination of environmental impact may not be feasible (Chien Bong et al., 2017).

Conclusion

Using red tilapia (*Oreochromis* Spp.) viscera chemical silage as an alternative protein substitute in laying hens' feed manufacturing offers several benefits. It allows for the utilization of fish by-products without requiring modifications to production parameters, while also improving feed conversion efficiency. Furthermore, exploiting fish by-products for laying hens' feed manufacturing reduces environmental load compared to conventional waste disposal processes such as landfill disposal, making it the most environmentally favorable production system.

However, it's important to note that the utilization of natural resources and wastewater disposal are the primary sources of adverse effects on the environment in this process. To mitigate these impacts, implementing renewable and efficient alternative energies can reduce the environmental load of the process while also creating products with commercial value. By adopting sustainable practices and technologies, it's possible to enhance both the environmental sustainability and economic viability of the production system.

Disclaimer

None

References

Ahmed, Z., & Wang, Z. (2019). Investigating the impact of human capital on the ecological footprint in India: An empirical analysis. *Environmental Science and Pollution Research,* 26(26), 26782–26796. https://doi.org/10.1007/s11356-019-05911-7.

Akinola, & Iyomo. (2018). Nigerian Society for Animal Production Nigerian Journal of Animal Production © Egg quality analysis and performance of laying hens fed different levels of calcium 1 1. *In Nig. J. Anim. Prod* (Vol. 45, Issue 1).

Alicorp. (2010). *Manual de Crianza Tilapia* (p. 49).

Batalha, O. de S., Alfaia, S. S., Cruz, F. G. G., Jesus, R. S., Rufino, J. P. F., & Silva, A. F. (2018). Pirarucu by-product acid silage meal in diets for commercial laying hens. *Revista Brasileira de Ciencia Avicola,* 20(2), 371–376. https://doi.org/10.1590/1806-9061-2017-0518.

Camaño, J. A., Rivera, A. M., & Zapata, J. E. (2020). Efecto del espesor de película y de la ubicación de la muestra en un secador solar directo, sobre la cinética de secado de ensilado de vísceras de tilapia roja (*Oreochromis* sp). *Información Tecnológica,* 31(1), 53–66. https://doi.org/10.4067/s0718-07642020000100053.

Chien Bong, C. P., Ho, W. S., Hashim, H., Lim, J. S., Ho, C. S., Peng Tan, W. S., & Lee, C. T. (2017). Review on the renewable energy and solid waste management policies towards biogas development in Malaysia. *Renewable and Sustainable Energy Reviews*, 70(December), 988–998. https://doi.org/10.1016/j.rser.2016.12.004.

Coorey, R., Novinda, A., Williams, H., & Jayasena, V. (2015). Omega-3 Fatty Acid Profile of Eggs from Laying Hens Fed Diets Supplemented with Chia, Fish Oil, and Flaxseed. *Journal of Food Science,* 80(1), S180–S187. https://doi.org/10.1111/1750-3841.12735.

FAO. (2018). The State of world fisheries and aquaculture. *Meeting the Sustainable Development Goals.* https://doi.org/Licencia: CC BY-NC-SA 3.0 IGO.

Farrell, D. (2013). *Función de las aves de corral en la nutrición humana.* Fao, 2.

Gaviria G, Y. S., Londoño F, L. F., & Zapata M, J. E. (2020). Effects of chemical silage of red tilapia viscera (*Oreochromis* spp.) as a source of protein on the productive and hematological parameters in isa-brown laying hens (Gallus gallus domesticus). *Heliyon,* 6(12). https://doi.org/10.1016/j.heliyon.2020.e05831.

Gwehenberger, G., & Narodoslawsky, M. (2007). The ecological impact of the sugar sector- Aspects of the change of a key industrial sector in Europe. *Computer Aided Chemical Engineering,* 24, 1029–1034. https://doi.org/10.1016/S1570-7946(07)80196-9.

Henriksson, P. J. G., Rico, A., Zhang, W., Ahmad-Al-Nahid, S., Newton, R., Phan, L. T., Zhang, Z., Jaithiang, J., Dao, H. M., Phu, T. M., Little, D. C., Murray, F. J., Satapornvanit, K., Liu, L., Liu, Q., Haque, M. M., Kruijssen, F., De Snoo, G. R., Heijungs, R., … Guinée, J. B. (2015). Comparison of Asian Aquaculture Products by Use of Statistically Supported Life Cycle Assessment. *Environmental Science and Technology,* 49(24), 14176–14183. https://doi.org/10.1021/acs.est.5b04634.

Ibidhi, R., Hoekstra, A. Y., Gerbens-Leenes, P. W., & Chouchane, H. (2017). Water, land and carbon footprints of sheep and chicken meat produced in Tunisia under different farming systems. *Ecological Indicators,* 77, 304–313. https://doi.org/10.1016/j.ecolind.2017.02.022.

IDEAM. (2011). Estimación de las emisiones de dióxido de carbono generadas por deforestación durante el periodo 2005-2010.

IPCC. (2006). Directrices para los inventarios nacionales de gases de efecto invernadero - *Elimimación de desechos sólidos - Capitulo* 3. In IPCC.

Kjos, N. P., Herstad, O., Skrede, A., & Øverland, M. (2001). Effects of dietary fish silage and fish fat on performance and egg quality of laying hens. *Canadian Journal of Animal Science,* 81(2), 245–251. https://doi.org/10.4141/A00-086.

Leinonen, I., Williams, A. G., Wiseman, J., Guy, J., & Kyriazakis, I. (2012a). Predicting the environmental impacts of chicken systems in the united kingdom through a life cycle assessment: Egg production systems. *Poultry Science,* 91(1), 26–40. https://doi.org/10.3382/ps.2011-01635.

Leinonen, I., Williams, A. G., Wiseman, J., Guy, J., & Kyriazakis, I. (2012b). Predicting the environmental impacts of chicken systems in the united kingdom through a life cycle assessment: Egg production systems. *Poultry Science,* 91(1), 8–25. https://doi.org/10.3382/ps.2011-01635.

Malakahmad, A., Abualqumboz, M. S., Kutty, S. R. M., & Abunama, T. J. (2017). Assessment of carbon footprint emissions and environmental concerns of solid waste treatment and disposal techniques; case study of Malaysia. *Waste Management,* 70, 282–292. https://doi.org/10.1016/j.wasman.2017.08.044.

Mamouni Limnios, E. A., Ghadouani, A., Schilizzi, S. G. M., & Mazzarol, T. (2009). Giving the consumer the choice: A methodology for Product Ecological Footprint calculation. *Ecological Economics,* 68(10), 2525–2534. https://doi.org/10.1016/j.ecolecon.2009.04.020.

Martínez-Alvarez, O., Chamorro, S., & Brenes, A. (2015). Protein hydrolysates from animal processing by-products as a source of bioactive molecules with interest in animal feeding: A review. *Food Research International,* 73(1069), 204–212. https://doi.org/10.1016/j.foodres.2015.04.005.

Padhi, M. K., Chatterjee, R. N., Haunshi, S., & Rajkumar, U. (2013). Effect of sulfonamides on egg quality in colour layers. *Indian Veterinary Journal,* 48–1(2), 122–125.

Perez López, D. D. J., Monroy, J. P., Reyes Ramírez, A. K., Huerta, A. G., & Sangermán Jarquín, D. Ma. (2019). Fertilización orgánica con tres niveles de gallinaza en cuatro cultivares de papa. *Revista Mexicana de Ciencias Agrícolas,* 10(5), 1139–1149. https://doi.org/10.29312/remexca.v10i5.1759.

Perez-Martinez, M. M., Noguerol, R., Casales, B. I., Lois, R., & Soto, B. (2018). Evaluation of environmental impact of two ready-to-eat canned meat products using Life Cycle Assessment. *Journal of Food Engineering,* 237(May), 118–127. https://doi.org/10.1016/j.jfoodeng.2018.05.031.

Quiroz Fernández, L. S., Izaquierod Kulich, E., & Menéndez Gutierrez, C. (2018). Estudio del impacto ambiental del vertimiento de aguas residuales sobre la capacidad de autodepuración del río Portoviejo, Ecuador. *Centro Azúcar,* 45(01), 73–83.

Ravindran, V. (2013). *Disponibilidad de piensos y nutrición de aves de corral en países en desarrollo - Alimentos alternativos para su uso en formulaciones de alimentos para aves de corral.* Fao, 4.

Rostagno, H. S., Texeira, L. F., Lopez, J., Cezar, P., Flávila de Oliveira, R., Clementino, D., Soares, A., Lluiz de Toledo, S., & Frederico, R. (2011). *Tablas Brasileñas Para Aves Y Cerdos.* 157–166.

Silva, A. F., Cruz, F. G. G., Rufino, J. P. F., Miller, W. M. P., Flor, N. S., & Assante, R. T. (2017). Fish by-product meal in diets for commercial laying hens. *Acta Scientiarum - Animal Sciences,* 39(3), 273–279. https://doi.org/10.4025/actascianimsci.v39i3.34102.

Skunca, D., Tomasevic, I., Nastasijevic, I., Tomovic, V., & Djekic, I. (2018). Life cycle assessment of the chicken meat chain. *Journal of Cleaner Production,* 184, 440–450. https://doi.org/10.1016/j.jclepro.2018.02.274.

Tiwari, A. (2016). A Review on Solar Drying of Agricultural Produce. *Journal of Food Processing & Technology,* 7(9). https://doi.org/10.4172/2157-7110.1000623.

Zhou, E., Pan, X., & Tian, X. (2008). Application Study of Xylo-oligosaccharide in Layer Production. *Modern Applied Science*, 3(1), 103–107. https://doi.org/10.5539/mas.v3n1p103.

Index

A

B

C

M

N

O

P

R

S

T

V

W